'Ivo Vegter's book is a breath of fresh air in a
ing stale. His style is clear without being cond
of nothing.'   *– Dominic*

former edi. . ., ... Junday Telegraph

'When truth is obscured by fear mongering and sensational, emotional
rhetoric, our ears fall deaf to reason. Ivo stands tall as he calls us to understand the painful legacy we will create because we allowed ourselves to be
manipulated by politicians and those well-intentioned individuals who
inadvertently play directly into the hands of those who seek to disenfran-
chise the people and enrich only themselves. Ivo exposes the lies one by
one, and calls for balance, a voice of reason in the wilderness.'
– Hugh (Bob) Glenister, who successfully challenged the dissolution of the
Directorate of Special Operations (the Scorpions) in the Constitutional Court

'A much-needed international voice on enviro-insanity. The voice we
were longing to hear. Vegter has already won recognition in the US as one
of the clearest in all of media. Heed Vegter.'
– Amity Shlaes, syndicated columnist for Bloomberg *and*
senior fellow at the Council on Foreign Relations

'There is something enormously patronising about regulators and environmentalists who make decisions on behalf of others about what's good
and bad for them. It assumes that they know more than others, and how
they interpret the world is more important than how others view the
world. Such actions are fundamentally against individual freedom. Ivo
Vegter dissects that logic with characteristic wit, verve and reason in this
absorbing account.'
– Salil Tripathi, independent journalist published in LiveMint, India Today,
the Far Eastern Economic Review, The Wall Street Journal *and the*
International Herald Tribune

EXTREME
ENVIRONMENT

How environmental exaggeration harms
emerging economies

Ivo Vegter

Published by Zebra Press
an imprint of Random House Struik (Pty) Ltd
Company Reg. No. 1966/003153/07
Wembley Square, First Floor, Solan Road, Gardens, Cape Town, 8001
PO Box 1144, Cape Town, 8000, South Africa

www.zebrapress.co.za

Published in 2012

1 3 5 7 9 10 8 6 4 2

Publication © Zebra Press 2012
Text © Ivo Vegter 2012

Cover images © Sundeip Arora, iamwahid, Billy Alexander and stock.xchng

PUBLISHER: Marlene Fryer
MANAGING EDITOR: Ronel Richter-Herbert
EDITOR: Lisa Compton
PROOFREADER: Ronel Richter-Herbert
COVER DESIGN: Michiel Botha
TEXT DESIGN: Jacques Kaiser
TYPESETTER: Monique van den Berg/ Jacques Kaiser

Set in Janson MT 11 pt on 15.5 pt

Printed and bound by Paarl Media
Jan van Riebeeck Drive, Paarl, South Africa

ISBN 978 1 77022 364 6 (print)
ISBN 978 1 77022 365 3 (ePub)
ISBN 978 1 77022 366 0 (PDF)

Contents

Acknowledgements vii

Foreword xi

Introduction – Exaggeration, Sensationalism and Age-old Neuroses . . . 1

1 David versus Goliath – The Rise of the Fracking Controversy 9
2 Fracking – A Light on Scary Nightmares 30
3 Fracking – Crying Wolf 53
4 *Umami* – The Good Taste 81
5 Many People Died, But Not at Fukushima 108
6 The Caribbean – Reports of My Death Were Greatly Exaggerated . . . 138
7 Fun with Facts and Fallacies 160
8 The Last Resort – Climate Change 192
9 The Church of Gaia 221
10 Exaggeration Makes Us Poorer 245

Endnotes 267
References 283
Index 313

For my late grandmother, after whom I was named, and who survived World War II only to live out her final years in mortal fear that her little house in Holland would be inundated by rising sea levels.
Rest in peace, *Oma* Vegter.

And for my parents, who raised me with an inquiring mind, a love for science, an appreciation of nature and respect for all humanity.

Acknowledgements

Much of the substance of this book derives from the regular columns I write for the *Daily Maverick*. I owe its editor and publisher, Branko Brkic, a debt of gratitude. He encouraged me to start writing opinion over a decade ago, when I was unconvinced that I had what it takes. He has generously supported me as the scope of my writing expanded over the years to embrace my interests in economics, science and history. He has given me assignments that broadened my experience of subjects with which I was unfamiliar and uncomfortable, which helped to develop my critical research and reasoning skills.

He has also been exemplary in giving me the freedom to express opinions with which he thoroughly disagrees, and has had the integrity and courage to shield me from even knowing who advertises in his publications, let alone when those advertisers place pressure on him, as they surely have done on occasion. Ours has been a long and productive association which I will always value.

There are many friends, family and acquaintances who have tolerated my enthusiasm for the subjects in this book, and have listened at length to discourses in which I examined my assumptions and field-tested my lines of argument. Many members of mailing lists and online forums of which I have been a member over the years have likewise helped to hone my arguments.

I have long had a policy of defending, in public, the arguments that I make. It is only right that I am expected to do so, rather than to simply publish a column and leave it to stand on its own, above the fray. Hundreds of readers have given of their time and energy to contribute in uncountable ways to the views I now hold. A few have become famous enemies, but even they have been sharp-eyed in spotting my errors, vicious in exposing weaknesses in my research or reasoning, and generous in adding substance to lines of argument that hadn't occurred to me.

There are too many people who have influenced my thinking over the years to name individually, but I'd like to make two exceptions. The first is Hugh Glenister, a tireless defender of freedom in South Africa, who generously sponsored a trip to FreedomFest 2011 in Las Vegas for five South Africans, among whom I was one. The opportunity to discuss economic liberty with some of the sharpest free-market minds in the world is one I will not soon forget, and has motivated much of the argument presented in this book. The second is an old friend and drinking buddy, Jeff Delaney. He has been an important influence in sharpening my critical thinking, daring me to take unpopular positions and demolishing my conceits. He has for many years served as a reliable sounding board and source of ideas, and to this day remains an undaunted and tireless critic of my unforgiveable prolixity. Delaney, hail to the chief.

To my publisher at Zebra Press (an imprint of Random House Struik), Marlene Fryer, who thought this book would be a valuable contribution to public discourse in South Africa and beyond, and generously offered the experience, marketing reach and publishing resources of the Zebra Press imprint to make it a reality, my thanks. To the managing editor, Ronel Richter-Herbert, my gratitude for your endless patience and determination to see this project through to its successful conclusion, despite every obstacle a shiftless and obstinate author could raise against it. And to my sharp-eyed editor, Lisa Compton, who caught many an error of both style and substance and was never shy to challenge my instincts and assumptions, my respect for improving this book in too many ways to mention. Any remaining oversights or misjudgements are all mine.

Finally, to all my friends in my recently adopted hometown of Knysna – the most beautiful town in the world, in a lovely natural setting about which I care more than some of my critics might imagine – my apologies for being so scarce. You've all been wonderfully supportive and enthusiastic about this project, even though many of you are really socialist eco-hippies who think I'm a secret agent for global free-market capitalism. All of that is true, except for the 'secret' part. Here's to you.

IVO VEGTER
JUNE 2012

Foreword

Catastrophism, lies and exaggerations by anti-liberty activists have been exposed and debunked repeatedly.

'Do we really need another voice of reason?' I pondered, when Ivo Vegter offered me his manuscript.

However, this book is truly different. Unlike earlier critics of alarmism, Ivo is willing to give passionate activists the benefit of any doubt, but bases his judgments on factual evidence. Ivo cares deeply about integrity, truth, the environment and, above all, humanity. He sifts skilfully through fact and fiction on controversial issues. He accepts the burden of proof that rests with anyone who questions the established view.

He provides a long overdue third-world perspective on environmentalism and health scares that impact the world's poor billions at the behest of its rich millions. The poor should not, he argues, be subjected to policies only the rich can afford.

He shows that far from merely being harmless, concerned individuals, green propagandists (rather than their adversaries) ignore inconvenient truths and are shamelessly selfish. After all, they seek to impose *their* preferences by law instead of by voluntary market choice, regardless of what misery they might inflict. For them to accuse 'greedy capitalists', who can profit only by serving others, of being uncaring is doubly perverse, especially in a country like South Africa, which is characterised by extreme unemployment and poverty.

Ivo laments and exposes the extent to which hard science and economics are subverted by self-serving officials, consultants, contractors and activists who are richly rewarded for toeing the line; they wallow in prestige, power, grants, subsidies and the like.

The book is replete with examples of Thomas Sowell's observation that 'Ego trips by coteries of self-exalting people are treated in the media as idealism, rather than the petty tyranny it is.'

Seemingly disparate issues are addressed in varying degrees of detail, starting with a comprehensive analysis of 'fracking'. Ivo tackles the hard questions, and we visit the Niger Delta, for instance, where Shell supposedly devastated a pristine paradise, as well as the Deepwater Horizon oil spill and the nuclear incident at Fukushima. We find that the price-mechanism counteracts the risk of excessive 'resource depletion', and that fears about cancer caused by food or chemicals are essentially baseless.

In short, this book isn't for incurable pessimists. Open-minded readers will be left with no doubt about the need for level-headed policymaking and critical evaluation of sensational claims from whatever source, especially in developing countries.

If things are as clear-cut as Ivo convinces us they are, why is bad news good news? Why does fabricated catastrophe trump truth? Hold forth at a dinner party on higher-than-ever crime, more-endangered-than-ever species, worse-than-ever disease, or any other 'the sky-is-falling' myth, and you're a hit. Tell them there's nothing to worry about, present hard facts, and you're ignored or vilified. What explains the predilection for alarmism and denialism of the fact that most things get better for most people most of the time in most places?

Is the media to blame? No, journalists supply what people want, and most people want to be told things are getting worse, because there's too much freedom.

Detractors typically attack the messenger, not the message. But Ivo is uninfluenced by special interests of any kind. This is not the work of an apologist, but a rare example of a formidable, enquiring and honest mind, willing to engage readers in his popular columns.

LEON LOUW

EXECUTIVE DIRECTOR OF THE FREE MARKET FOUNDATION OF SOUTH AFRICA

11 JULY 2012

'That man is prudent who neither hopes nor fears anything from the uncertain events of the future' — Anatole France

'A thing is not proved just because no one has ever questioned it. What has never been gone into impartially has never been properly gone into. Hence scepticism is the first step toward truth' — Denis Diderot

Introduction
Exaggeration, Sensationalism and Age-old Neuroses

The fear of new things – new drugs, new technology, new industries, new lifestyles – is deep-rooted. You can see it around the dinner table or in the pub, whenever people denounce some drug, some food additive or some local industrial development. You can see it surfacing in the history books over the centuries, in many guises.

One such event – activism and industrial sabotage by textile artisans in the early nineteenth century – gave rise to the term 'Luddite', which in modern usage means an aversion to new technology. In fact, the followers of the mythical Ned Lud weren't simplistically opposed to modernity. They were not unique, nor were they even the first – or last – of their kind. Their fears, like the fears of many prophets and subversives before them, were complex and deeply rooted in economic displacement and disruption. Both foreign competition and retaliatory protectionism hurt the revenues of the industries that employed them. Rising food prices squeezed how far their money would stretch. But the focus of their anger became the industrial-era machinery that increased the productivity and profitability of textile mills while reducing wages and enabling lower-skilled workers, like apprentices and even women (oh, the iniquity!), to displace the highly skilled artisans their trade formerly required.

History is replete with examples of legislation designed to protect established industries, and to raise barriers to entry to new ones. In seeking to prevent disruption and apparently unfair turns of fortune for some, these regulations have always had a stultifying, conservative effect on countries, curbing their economic vitality and hindering technological progress.

In many ways, environmentalism and health regulation play a similar role today. Under the guise of protecting people and the environment from the negative effects of development, they conveniently protect established industries while making it hard for newer, more efficient competition to arise. The consequences sometimes – often, even – are worse than the ills these regulations were designed to avert.

Imagine if modern environmental regulation had been around when the horseless carriage, the coal-fired power station, the fertiliser industry, the light bulb or computer electronics had been invented. Electronics contain heavy metals and other toxic chemicals that pose a challenge for disposal and are released into the environment during accidents such as fires. Fertiliser is believed to be 'unnatural' (in contrast to some vague, ill-defined, emotional concept of 'natural') and leaches into groundwater and streams, where it causes dangerous algae blooms that suck the life out of the local ecosystem. Power stations are horrible polluters, even before we take into account the carbon dioxide emissions that are alleged to raise the temperature of the planet. The light bulb may be cheap and wonderfully effective, but it is a relatively inefficient user of energy. Ironically, what once was the universal symbol of innovation and ideas has recently become the subject of disapproving looks and outright bans. And then there is the car. That invention would never have passed modern safety and environmental rules. It kills defenceless pedestrians and mutilates innocent passengers, while emitting unhealthy smog into the atmosphere and running on a toxic cocktail of petrochemicals and additives like ethylene glycol.

While the rich world can arguably afford to be fussier nowadays than it was during its own development – and all indications are that prosperity is positively correlated with greater care for the environment – large parts of the world have yet to achieve that state of discerning wealth. In the rich world, heavy industrial regulation merely makes rich people a little poorer. It may seem to them a fair trade-off for a safer workplace and a healthier, more pleasant environment. In the poor world, however, the regulatory burden causes unemployment, disease and deprivation.

That is not to say we ought not to learn from mistakes. It doesn't mean we shouldn't be sensible about how we develop economically and how we utilise our environment productively. It is clear that some activities – notably in areas where private property rights are difficult to establish, such as fisheries and forests – are unsustainable, and others pose clear risks to human health and well-being.

However, to make rational decisions that enable the poor to prosper, as they have done wherever people have enjoyed economic freedom, we need accurate information. Exaggerating environmental and health fears may indeed prompt regulatory action, which by chance may turn out to be beneficial. But it also ensures that regulation too often goes further than it needs to, and prevents trade-offs that poor people would be happy to make in order to improve their lives.

The cost of environmental regulation is particularly acute in developing countries, and this is where the insistence on upholding rich-world standards, not to mention the outright lies and exaggeration that support this insistence, has the most harmful impact.

In this book, we'll look at some examples of how environmental activists exaggerate their fears to promote a perversely conservative and oppressive form of government. We'll analyse their claims and examine to what extent they're true, and to what extent they're just fear-mongering aimed at stifling economic activity. We'll speculate about motives, which range from merely fashionable health neuroses to ideological opposition to freedom, industry and capitalism.

These fears and exaggerations are nothing new, but the extent to which they find expression in restrictive and costly regulation is a modern phenomenon. It is a feature of modern government that sometimes improves lives and human health, but often imposes burdens on society that make its overall condition worse rather than better.

To evaluate claims of environmental destruction, atmospheric pollution or toxic threats to human health, we need not only an economic outlook that prizes the freedom that has produced so much productivity and prosperity – we also need facts, not exaggeration or sensationalism.

It is not good enough that a prominent speaker on corporate ethics, such as South Africa's Justice Mervyn King, famous for the King Report on Corporate Governance, says that the world may be facing an eight-metre rise in sea levels by 2100, when not even the worst-case scenarios of the most extreme environmentalists support such a wild claim. Companies cannot make rational cost-benefit decisions in the face of grossly exaggerated assertions of risk, and those that do act on such misinformation cause severe economic harm that affects not only their shareholders, but also their customers, their employees and, ultimately, society at large.

It is not good enough that environmental polemic and propaganda peddled in so-called documentary films – such as *Gasland*, Josh Fox's award-winning screed against natural gas drilling –

influences government decisions when the claims in such films are questionable at best. Most claims by the environmental groups opposed to the technique known as hydraulic fracturing turn out, upon closer examination, to be exaggerations or even patently untrue. It might suit comfortable middle-class activists to stoke these fears in the pursuit of an ideology that opposes any economic progress and industrial development that has even the least impact on our environment. However, it ought not to be the basis for policymaking in a country like South Africa, where even official unemployment numbers still stick stubbornly near 25 per cent, and where millions continue to live in abject poverty almost two decades after their liberation from racial oppression.

The media is often complicit in the exaggeration campaigns of environmental activists. Journalists are naturally sympathetic to narratives that purport to speak truth to power, whether that power is political or economic. The image of a valiant David fighting a corporate Goliath is extremely seductive. Sensationalist headlines are always tempting to editors, and too often displace more sensible analyses that question how serious, widespread or true popular alarmism really is. Emotional environmentalism, whether online, on television or in glossy magazines, tends to trump dull statistics and economic theory.

As a consequence, news events such as the damage to the Japanese nuclear power station at Fukushima get portrayed as images of unmitigated catastrophe. Thousands of people died in the natural disaster that caused the accident, while no one has (yet) died as a result of radiation exposure at Fukushima. Instead of considering it a testament to the safety of nuclear power that even a very old reactor design under less-than-sterling management did not cause far worse damage under such extraordinarily severe circumstances, the sensational storyline of nuclear meltdown trumps even the mind-numbing reality of 20 000 dead or missing people. The nuclear fear, duly whipped up by special-interest lobby groups such as Greenpeace, causes both government policy-makers and ordinary people to distrust a source of energy that has over time proved to be safer than any other alternative, economically viable or not.

The same storyline plays out time and time again. Exaggerated and misplaced fears about pollution or human health hamper the development of more abundant and cheaper energy resources, increase the cost of food and put medical care out of reach of the poor. More broadly, they retard technological development, impose unwarranted costs on industry and raise prices for everyone. Environmental exaggeration has a direct and negative impact on economic prosperity. While rich activists might be able to afford fashionable green technology and the subsidies that they invariably require, the world's remaining poor population cannot indulge such luxuries. Their concerns are more basic, and their choices more stark.

This book begins with an examination of the claims of the 'anti-fracking movement', which opposes shale-gas extraction in South Africa. It reveals most of their claims to be exaggerated or false. It looks into similar fears that affect our food, energy and agricultural industries, where all sorts of modern improvements are – sometimes justifiably, but often not – suspected of causing severe impacts, such as cancer. To make the more general case, the book examines the exaggerations perpetrated by the environmental movement even in very real cases of disaster, such as the Deepwater Horizon oil spill and the Fukushima nuclear reactor meltdown. It then puts all this environmental extremism in the context of motives both ideological and psychological, and considers their possible impacts on developing nations such as South Africa.

Throughout, the point is not to say that all environmental warnings are false. Some are clearly true, and some new chemicals, techniques, products or industrial processes really ought to be viewed with suspicion, or even advised against. Greedy and reckless industrialists may not be as common as the caricature would have it, but they surely do exist and merit our concern.

However, when the activists that claim to look out for the interests of ordinary people make exaggerated or even false claims about the dangers of development, they muddy the waters for policymakers. Ironically, they undermine their own credibility by crying wolf, which means that their valid concerns get ignored. Who isn't tired of constant whining about how margarine will save your life, or kill you, depending on who you ask and when you ask them? People can be forgiven for scepticism about the knee-jerk anti-capitalism of environmental extremists, and their routine opposition to any and all industrial development or use of cost-effective sources of energy.

More dangerously, however, the inaccurate information that is spread by environmental activists and their enablers in the sensationalist media has very real impacts on regulation and policymaking. When burdensome rules are made that prove to be unnecessary or excessive, those rules serve only to drive up the cost of industrial production. True, this affects the profits of multinationals, about which nobody other than supposedly rich shareholders cares, but it also endangers employment, and drives up prices for the essentials of a decent life.

Environmentalists like to claim that we ought to be cautious, and that precaution never did any harm. This is as false as the notion that reckless risk taking can't cause any harm. It is certainly true that excessive risk can have all sorts of really harmful consequences. But excessive caution, prompted by exaggerated fears and ideological opposition to development, can be just as dangerous.

When poor people can't find jobs because companies can't afford to hire more people or are prevented from exploiting new opportunities, environmental exaggeration has very real human suffering to answer for. When people can't afford food because farmers and food production companies are required to jump through unnecessary hoops just to get produce to market, irrational fears cause very real malnutrition and childhood development risks.

All life is a trade-off between costs and benefits. All production to furnish both the essentials and the luxuries of life consists of taking informed risks and working hard to make sure those risks prove worth taking. However, to evaluate costs against benefits, and to mitigate risk, we need accurate information, sound data and thoughtful analysis on which to base our decisions.

Exaggerating benefits while ignoring risks is obviously naive, and many a cautionary tale has been written about it. My hope is that this book will serve as a cautionary tale about the opposite extreme: that it will highlight, especially for developing nations, the dangers of exaggerating risks while ignoring the benefits of economic liberty and industrial development. The notion that the environment can be left untouched by human hands is absurd. Our historical imposition on nature to farm our food, build our homes, lay out our roads and fence our properties has created a duty of actively managing, carefully protecting and responsibly exploiting natural resources. More importantly, humanity still needs a healthy and productive environment, if only to raise the living standards of those countries that have yet to reach the longevity, health and prosperity of the developed world.

1

David versus Goliath

The Rise of the Fracking Controversy

THE SPEECH

At first they were under the radar. Only a handful of readers of specialist oil-industry publications were even aware that several domestic and foreign oil companies were conducting preliminary studies into potential shale-gas resources in South Africa.

Unlike in the US, where landowners stand to earn a 12.5 per cent royalty for gas finds on their properties, South Africa's mineral rights were nationalised in 2002. As a result, discussions between oil companies and the government agency responsible for petroleum resources took place in boardrooms far removed from the land under discussion, and out of sight of the media and the general public.

In March 2010, the veteran publisher of *Mining Weekly*, Martin Creamer, reported that a number of companies and consortia, having completed their earlier technical investigations, had applied for formal rights to explore for gas by drilling in the Karoo Basin.

What they're after lies buried by the ages, several kilometres deep in a layer of rock known as Ecca shale. The Karoo Basin extends further, in geological terms, than what mere surface dwellers would recognise as the barren but hauntingly beautiful semi-desert of scrubland, supporting widely dispersed flocks of hardy sheep. It covers some 600 000 km², or about half of South Africa's land area, and the shale under it is believed to contain potentially vast reserves of methane, or natural gas. To get to it, energy companies propose to use a technique called 'hydraulic fracturing', in which liquid under pressure is used to crack the rock surrounding a well. Among mining and drilling engineers, this is known as 'fracking' for short.

In the US, hydraulic fracturing – a technique that has been used successfully for over sixty years to extend the life of fossil-fuel reservoirs and reach oil and gas trapped in rock formations – has been combined with modern drilling technology, which permits well shafts to be sunk to a great depth before drilling continues at an angle, or even horizontally. The advances in these two techniques have sparked a dramatic shale-gas boom.

Although the Karoo Basin is by no means the only promising oil and gas region in South Africa and its territorial waters, interest in the Karoo's potentially vast shale deposits rose sharply as the shale-gas boom took off in the US and elsewhere. Initial estimates rank it as the fifth-largest gas resource in the world, after China, the US, Argentina and Mexico.

The proposals under consideration by the South African government were that, in their search for natural gas, these drilling companies would use hydraulic fracturing first in exploration and then, if the reserves they found proved to be plentiful, in full production.

Enter Lewis Gordon Pugh, OIG. The mainstream media in South Africa had been blithely ignoring what appeared to be just another dull mining venture until Pugh, a well-known environmental campaigner and an ambassador for the World Wide Fund for Nature, gave a rousing anti-fracking speech in Cape Town on 25 March 2011. Pugh is the son of a nurse and a navy admiral, who reportedly had been deeply impressed by the post-war fears of nuclear testing and the rise of the environmental movement. Pugh himself rose to fame in recent years for long-distance swimming, particularly in extremely cold water. The OIG behind his name stands for the Order of Ikhamanga (Gold), which is awarded by South Africa for exceptional achievement in the arts, culture, literature, music, journalism or sport.

A number of Pugh's exploits, like swimming the length of the Thames and the width of the Maldives, were designed to raise awareness about climate change. His most astonishing sporting achievement, however, was a one-kilometre swim in -1.7 °C water, along a temporary crack in the ice at the North Pole during the summer of 2007. With this celebrity status to back him, the respected environmentalist proceeded to speak about gas drilling in the Karoo. One of the firms that applied for exploration permits was Shell, and that made a sufficiently juicy target to rouse Pugh's interest.

But Shell wasn't the only company to apply. The first exploration permit was applied for by Bundu Gas and Oil, a subsidiary of the Australian firm Sunset Energy, which specialises in deep-shale resources. Fresh from successes in the Barnett Shale formation in the state of Texas, the US exploration company Falcon Oil & Gas followed hot on Bundu's heels. Shell International came in third. Between them, they seek to explore a broad strip of the Karoo, stretching from Queenstown in the east much of the way across the Karoo to Calvinia in the west, and from a line between Saldanha Bay and East London in the south, northwards towards Victoria West and Carnarvon.

In addition, the South African specialist in producing liquid fuels from coal and gas, Sasol, teamed up with Norway's Statoil and Chesapeake Energy, well known in the Marcellus Shale region of the north-eastern US, to obtain a technical cooperation permit, which allows prospecting without exploratory surface operations, for a large region to the north-west of Lesotho. It has since opted not to pursue an exploration permit. Another firm, Anglo Operations, is prospecting south-east of the Drakensberg, on the KwaZulu-Natal side, where geology suggests oil shale might be found, while an obscure group known as Moonstone Investments 81 is prospecting near the Kgalagadi Transfrontier Park in the Kalahari region of the Northern Cape. Figure 2 in the colour section of this book shows petroleum exploration projects currently under way in South Africa.

When readers of my columns in the *Daily Maverick* drew my attention to warnings that energy companies were preparing to 'frack the Karoo', my first instinct, informed by Pugh's speech and the popular press, was to oppose it. After all, without knowing much about it, the thought of cracking rock under pressure, using liquids that could end up anywhere, sounded remarkably harmful to the surrounding environment, and particularly to the groundwater. Even if you believe that economic development and environmental protection require reasonable trade-offs and compromises, Lewis Pugh made fracking the Karoo sound like a destructive industrial process that could cause serious harm to the people who live and farm on this land.

Because most of the media and the general public picked up on the fracking story only after Pugh's speech, and because it is a model of high environmental rhetoric, it is worth reproducing in full here. At the time of writing, the speech remains the headline item on Pugh's personal website.[1]

Standing Up to Goliath

25 March 2011
by Lewis Pugh
Cape Town, South Africa

I want to take you back to the early 1990s in this country. Nelson Mandela had been released. There was euphoria in the air. However, there was also widespread violence and deep fear.

This country teetered on the brink of a civil war. But somehow we averted it. It was a miracle!

And it happened because we had incredible leaders. Leaders who sought calm. Leaders who had vision. So in spite of all the violence, they sat down and negotiated a new Constitution.

I will never forget holding the Constitution in my hands for the first time.

I was a young law student at the University of Cape Town. The Constitution was the cement that brought peace to our land. It held our country together. The rights contained in it made us one.

I remember thinking to myself: never again will the rights of South Africans be trampled upon. Now every one of us – every man and every woman, black, white, coloured, Indian, believer and non-believer – has the right to vote. We all have the right to life. And our children have the right to a basic education. These rights are enshrined in our Constitution.

These rights were the dreams of Oliver Tambo. These rights were the dreams of Nelson Mandela. These rights were the dreams of Mahatma Gandhi, of Desmond Tutu and of Molly Blackburn. These rights were our dreams.

People fought – and died – so that we could enjoy these rights today.

Also enshrined in our Constitution is the right to a healthy environment and the right to water. Our Constitution states that we have the right to have our environment protected for the benefit of our generation and for the benefit of future generations.

Fellow South Africans, let us not dishonour these rights. Let us not dishonour those men and women who fought and died for these rights.

Let us not allow corporate greed to disrespect our Constitution and desecrate our environment.

Never, ever did I think that there would be a debate in this arid country about which was more important – gas or water. We can survive without gas. We cannot live without water. If we damage our limited water supply – and fracking will do just that – we will have conflict again here in South Africa.

Look around the world. Wherever you damage the environment, you have conflict.

Fellow South Africans, we have had enough conflict in this land – now is the time for peace.

A few months ago I gave a speech with the former president of Costa Rica. Afterwards I asked him: 'Mr President, how do you balance the demands of development against the need to protect the environment?' He looked at me and said: 'It is not a balancing act. It is a simple business decision. If we cut down our forests in Costa Rica to satisfy a timber company, what will be left for our future?'

But he pointed out: 'It is also a moral decision. It would be morally wrong to chop down our forests and leave nothing for my children and my grandchildren.'

Ladies and gentlemen, that is what is at stake here today: Our children's future. And that of our children's children.

There may be gas beneath our ground in the Karoo. But are we prepared to destroy our environment for five to 10 years' worth of fossil fuel and further damage our climate?

Yes, people will be employed – but for a short while. And when the drilling is over, and Shell have packed their bags and disappeared, then what? Who will be there to clean up? And what jobs will our children be able to eke out?

Now Shell will tell you that their intentions are honourable. That fracking in the Karoo will not damage our environment. That they will not contaminate our precious water. That they will bring jobs to South Africa. That gas is clean and green. And that they will help secure our energy supplies.

When I hear this, I have one burning question. Why should we trust them? Africa is to Shell what the Gulf of Mexico is to BP.

Shell, you have a shocking record here in Africa. Just look at your operations in Nigeria.

You have spilt more than nine million barrels of crude oil into the Niger Delta. That's twice the amount of oil that BP spilt into the Gulf of Mexico. You were found guilty of bribing Nigerian officials – and to make the case go away in the US, you paid an admission of guilt fine of $48 million (R328m).

And to top it all, you stand accused of being complicit in the execution of Nigeria's leading environmental campaigner, Ken Saro-Wira [sic], and eight other activists. If you were innocent, why did you pay $15.5 million to the widows and children to settle the case out of court?

Shell, the path you want to take us down is not sustainable.

I have visited the Arctic for seven summers in a row. I have seen the tundra thawing.

I have seen the retreating glaciers. And I have seen the melting sea ice.

And I have seen the impact of global warming from the Himalayas all the way down to the low-lying Maldive Islands.

Wherever I go – I see it.

Now is the time for change. We cannot drill our way out of the energy crisis. The era of fossil fuels is over. We must invest in renewable energy. And we must not delay.

Shell, we look to the north of our continent and we see how people got tired of political tyranny. We have watched as despots, who have ruled ruthlessly year after year, have been toppled in a matter of weeks.

We too are tired. Tired of corporate tyranny. Tired of your short-term, unsustainable practices.

We watched as Dr Ian Player, a game ranger from Natal, and his friends took on Rio Tinto, one of the biggest mining companies in the world, and won.

And we watched as young activists from across Europe brought you down to your knees, when you tried to dump an enormous oil rig into the North Sea.

Shell, we do not want our Karoo to become another Niger Delta.

Do not underestimate us. Goliath can be brought down.

We are proud of what we have achieved in this young democracy – and we are not about to let your company come in and destroy it.

So let this be a call to arms to everyone across South Africa, who is sitting in the shadow of Goliath: Stand up and demand these fundamental human rights promised to you by our Constitution.

Use your voices – tweet, blog, petition, rally the weight of your neighbours and of people in power.

Let us speak out from every hilltop. Let us not go quietly into this bleak future.

Let me end off by saying this: you have lit a fire in our bellies, which no man or woman can extinguish.

And if we need to, we will take this fight all the way from your petrol pumps to the very highest court in this land. We will take this fight from the farms and towns of the Karoo to the streets of London and Amsterdam.

And we will take this fight to every one of your shareholders.

And I have no doubt that in the end, good will triumph over evil.

This is grand oratory. These are rousing words. They raise goosebumps even among hardened cynics. But whenever you see strong, emotive language like this, the smartest response is to raise an eyebrow. Facts speak for themselves. If a story is thin on facts and heavy on thinly sourced propaganda, that's when grand rhetorical style is needed to compensate for the lack of substance.

Remember when you were young and something exciting or scary happened? You'd rush to the nearest adult to tell the story, tripping over your breathless superlatives. Now recall how the grown-ups told you to calm down and tell them just the facts. That was how I reacted to Pugh's speech.

The headline alone should give one pause. Whenever someone claims to be a David fighting against a Goliath, the implied presumption is that the little guy is right. Not that Pugh is really such a little guy, even in the figurative sense. Not everyone gets decorated by the president, and as an ambassador for the World Wide Fund for Nature, he is part of a large, well-funded and highly organised lobby. Other organisations, such as Greenpeace and Earthlife Africa, have risen to his banner, and the lobby group that formed around this campaign, the Treasure the Karoo Action Group (TKAG), said it wants to amass a war chest of R10 million or more.

But even if Pugh were a little guy, it isn't a given that David is right and Goliath is wrong. Even the biblical David could be cast as an invader who set upon Goliath's camp and made war, pillaging and murdering the fleeing survivors. Perhaps he was justified in doing so – he certainly claimed his God's blessing – but the presumption of Goliath's cruelty and guilt is a tale written by the victor. The truth, as with any religion-fuelled conflict over land and resources, probably lies somewhere in between.

Likewise, just because Pugh and his well-intended band of protesters are taking on big oil companies doesn't mean that they're automatically right. They might be, but many people who challenge large companies are taking chances to advance their own special interests, or are merely misguided despite their good intentions. That is why courts and commissions exist to hear such cases and judge them on factual merit, rather than assuming that the little guy is always right.

But back to the speech. Having established himself in his audience's mind as a worthy crusader for the greater good, Pugh starts the show. From the outset, he casts the fight in grand historical terms. It is, to his mind, a struggle worthy of heroes. He invokes icons like Gandhi, Mandela, and the courageous men and women who fought oppression and were prepared to die for our freedom. This freedom, he says, is now under threat.

However, even if you detest fossil fuels and believe renewable energy to be the future, as Pugh clearly does, comparing some mining activity in a sparsely populated region to the injustice of systematic state violence against an oppressed majority is a very big statement. Gas drilling is hardly apartheid. Such an extraordinary claim requires extraordinary evidence.

Pugh tugs at the heartstrings, appealing to our concern for the future of our children's children. The premise is that we ought not to affect the environment at all, lest the world that future generations inherit is somehow worse than the one we have today. This is a curious position to take, when the rise of twentieth-century prosperity has made responsible environmental management a luxury many can now afford. While the early history of industrialisation involved too much pollution and disease, post-industrial economies are much better off in environmental and health terms precisely because they are prosperous.

Would it have made sense to object to highways or railroads on the basis that they'd be scars on the landscape? Would South Africans have enjoyed a better standard of living if environmentalists a generation or two ago had successfully opposed mining and industrialisation? Would we have any confidence in the technical knowledge and foresight of our grandparents, living fifty or a hundred years ago, to solve the problems we face in the modern world? Imagine explaining catalytic converters to a Roaring Twenties automobilist.

Pugh's appeal to our children's future is almost entirely emotional rather than rational.

He paints a broad-brush picture of climate change, asking us merely to accept his anecdotal observations as scientific truth. Assuming, for the sake of argument, that climate change is largely man-made and can be remedied by our future actions, a question arises: Would natural gas make matters worse or better? This is a question that ought to be answered before one can take an informed position in favour of or in opposition to gas drilling.

Soaring ever higher, Pugh's speech cleverly juxtaposes the political tyrants that at the time were fast falling to popular uprisings, implying that what he calls 'corporate tyranny' is somehow similar. The fall of dictators who had ruled with iron fists for decades, enriching themselves at the expense of their cowed populations, appeals to the moral sensibilities of his intended audience. Pugh speaks directly to the reflexive anti-capitalist sentiment that seems to pervade so much of the media, academia, and readers of Naomi Klein and Michael Moore. By drawing a parallel between political dictators and companies, he hopes to imply that these evil corporations will also get rich by using violence against defenceless citizens. But is this true?

Unlike dictators, companies are subject to both criminal and civil law, and can be charged with crimes or sued for damages if they abuse their own freedoms in order to infringe on the freedoms of others. As long as crime and corruption are not tolerated, the fact that they occasionally happen does not warrant taking away the freedoms of those who do abide by the law.

In fact, far from threatening to tear up South Africa's Constitution, companies are composed of individuals who are subject both to the rule of law and to the voluntary patronage of their customers. As we'll see, the right to a healthy environment and clean drinking water, while important, is unlikely to come under any particular threat from gas exploration in the Karoo.

Whether the oil companies involved will indeed play by the rules is, of course, a valid concern. The reputation of oil companies in general is none too shiny, and Shell's record in Nigeria is hard to defend, as Pugh observes. But is Shell's reputation deserved? If so, is oil in Nigeria really comparable to gas in South Africa? And given what we know, should we therefore reject any future gas exploration?

Comparing oil companies to the likes of Hosni Mubarak and Muammar Gaddafi might make for artful political rhetoric, but it does not meet the standards of scientific or legal argument on which policymakers ought to base decisions. As a lawyer himself, Pugh is undoubtedly aware of this.

His most alarming paragraphs warn of social unrest and perhaps civil war, sparked by conflict over South Africa's resources, particularly water. By casting the issue as a stark choice between natural gas and drinkable water, Pugh is claiming that these are mutually exclusive. That is a false dichotomy, as we shall see.

It's like saying that you can forswear all fat and be an ultra-fit endurance swimmer like Pugh, or you can eat red meat and be a corpulent wastrel with heart disease. True, those two extremes are conceivable, but neither is very likely in the ordinary course of events. You can be perfectly healthy while enjoying a nice steak once in a while, provided you don't overdo it and you take a few sensible precautions. One does not preclude the other.

So it is with natural gas drilling. If there are risks, are they large and unavoidable, or are they small and manageable? And what about the benefits? Are they worth the risk? Pugh's emotive exaggeration is designed to avoid such debate. It is explicitly aimed at banning any and all drilling for gas in South Africa. He hopes that enough people will side with him to convince the government not to take advantage of these apparently abundant resources.

The question is, can South Africans afford to indulge the apocalyptic fears that Pugh peddles in pursuit of an untouched – and underutilised – environment? To find out, I set out to examine the claims of the opponents of shale-gas drilling. I pulled together all the major issues raised by the TKAG, which formed around Pugh under the coordination of a nature photographer named Jonathan Deal, as well as claims from other sources. The documentary film *Gasland*, made by independent director Josh Fox, for example, has been widely screened in South Africa and around the world. Other opposition groups in regions where gas drilling by means of hydraulic fracturing occurs have also contributed to the public controversy, as have dozens of readers in hundreds of comments on columns I have written on the subject.

It bears reiterating that my initial reaction was to give the opposition argument the benefit of the doubt. I don't trust big oil companies – or indeed any companies that lobby governments for permits and permissions – any more than Lewis Pugh does. In fact, I don't trust big oil companies any more than I trust Lewis Pugh.

For this reason, I gave gas-drilling companies' public relations departments a wide berth. Only when I needed very specific information that could not be obtained elsewhere did I source it from the industry itself. I also avoided papers written by lobby groups known to be funded by the oil industry, such as Energy In Depth. I wanted to make sure not only that corporate spin wouldn't influence my view, but also that my conclusions could be seen to be independent. By self-imposed rule, I limited myself to information gleaned from regulators in various countries, technical knowledge from training institutions, such as the International School of Well Drilling, and various pieces of academic research.

Beyond extensive research and almost two decades of experience as a journalist, evaluating facts and arguments in fields as varied as technology and economics, I make no claims to specialist credentials in any particular area of the energy business or the numerous scientific fields – from geology to chemistry to hydrology – that relate to it. When lobbyists such as the TKAG raise issues specifically to influence public opinion, it usually doesn't (and probably shouldn't) require more than calm, rational thought to evaluate it. That, after all, is what most members of the public, and indeed most public policymakers, will devote to an issue such as this.

Here's what I found by the light of research and rational thought, contrary to my expectations: almost every major point of opposition to fracking turned out to be either wildly exaggerated or entirely false.

Let's examine each of these points one by one. In each case, the goal is to determine to what extent the concerns are valid, and how much is incorrect or is an exaggeration, attributable to the environmentalists who oppose fracking in the Karoo or anywhere else shale gas exists. But first, a brief primer on what exactly hydraulic fracturing is, and why this technology is used with shale-gas wells.

WHAT IS THIS FRACKING STUFF ANYWAY?

How does hydraulic fracturing happen, and why is it useful? If you're going to create cracks in rocks below the surface, doesn't it stand to reason that you'll risk trouble with water aquifers, and perhaps even cause earthquakes?

The fracking process is not hard to understand, but it is important to have a clear picture of how it works. It is easy, for example, to suggest that the cracks created by fracturing could reach to underground water aquifers and cause pollution, but this is technically about as likely as hitting a cow in a field while flying an aeroplane.

For many years, only vertical drilling for oil and gas was possible. In the case of natural gas, this limited drilling companies to pockets of gas that had made their way out of the deep rock where they originate. Rock layers typically lie horizontally, which makes vertical drilling a problem. Although drillers could theoretically reach those deep shales known to be full of natural gas, and although hydraulic fracturing has been a tried-and-tested technique for many decades and has been used to get more oil and gas out of over a million existing wells, the cracks created by fracturing simply don't extend far enough to tap significant areas of shale or tight sand from a single well. The number of shafts that would have to be sunk very close together to turn a shale bed into a productive gas field are simply too high.

New drilling techniques in which the head of the drill can be turned to continue drilling at an angle, or even horizontally, changed the game. At sea, this permitted more oil reserves to be reached from the same drilling platform. On land, it made it possible to drill a shaft along a length of gas-bearing rock or tight sand, so that each well head can reach far more subsurface gas.

The shale layers targeted by gas exploration companies in the Karoo lie at a depth of between 2.5 kilometres and 4 kilometres, in Ecca Group shale dating to the Lower Permian age, some 280 million years ago. The question is, what lies between the surface and those deep shales, and does any of it pose a major risk of groundwater pollution?

As a number of critical reports on Shell's environmental man-agement plan suggest, more detailed work is required to form a complete geohydrological picture of the proposed Karoo exploration area. However, it is possible to get a fairly good idea of the under-ground world of the Karoo from academic reports, existing water wells, five test wells sunk before 1970, when Mossgas exploration was taking place, and experience from similar geological formations elsewhere.

Freshwater aquifers usually lie up to about 500 metres deep. In the Karoo, indications are that this figure is probably closer to 300 metres. Ninety-six per cent of all Karoo water wells, serving towns and farms, are less than 100 metres deep.

Depending on the specific geology and proximity to sea level, brackish brine aquifers normally lie some distance below the freshwater aquifers. The brine isn't good for much, other than – ironically – hydraulic fracturing. Inland, brine aquifers tend to lie quite deep, though near sea level they get closer to the sur-face. According to Maarten J. de Wit, a researcher at the Africa Earth Observatory Network of the Nelson Mandela Metropolitan

University in Port Elizabeth, 'brackwater reservoirs exist at deep levels in the Karoo, and our recent … work has located large salt-water reservoirs at a depth greater than 1 kilometre'.[2]

Because brine is heavier than fresh water, the two don't mix. Freshwater aquifers float above brackish-water aquifers, even when the two are very close together, as they are at the coast. Anyone who has drilled a borehole or water spike near sea level will be aware that one can get fresh water by not drilling too deep. Lower down is where the salt water lies. Further inland, at higher elevations, hitting brackish water is typically only a risk when drilling exceptionally deep artesian water wells – as opposed to the more common shallow wells that only need to reach the water table. The gas-bearing shale lies much further down than even these brine aquifers, which in the Karoo lie very deep indeed.

The methods used for drilling wells differ, depending on the depth and specific geology. There is a surprising amount of study material available online, although I cannot recommend it as light bedtime reading. Subject matter ranges from the history of well drilling to detailed descriptions of techniques needed for specific well types.

Relatively simple drilling techniques and well construction are sufficient for French drains (sewerage disposal wells in areas without piped sewerage services) or shallow water-table wells, but things begin to get complicated with deeper artesian boreholes for municipal water supplies. Drilling engineers are also employed in the environmental sector, for jobs such as groundwater management. They are able to design wells and pumping systems to recharge depleted groundwater aquifers or even to rehabilitate polluted groundwater.

By the time you get to geothermal energy wells, oil wells, deep-injection waste-disposal wells or carbon-sequestration wells, you're dealing with highly sophisticated projects performed by specialist engineering firms in the mining and energy industry. Shale-gas wells also lie much deeper than any aquifer-related wells, and compare readily with geothermal and deep-injection wells in terms of drilling techniques and well construction.

To get to the shale-rock layer, gas companies would begin by drilling a well straight down until the level of the shale is reached. How exactly the hole is made depends on the composition of the soil or rock through which the well is bored.

The well is encased in a thick casing consisting of sleeves made of steel and other impervious materials, and reinforced with concrete. The technology for doing this has improved over the years, and wells are usually strong enough to withstand very high pressures and even significant seismic activity without rupturing.

The drill head is suspended from the surface and can be turned by an operator in order to drill a horizontal shaft radiating out from the bottom of the vertical well. One well can accommodate several such horizontal shafts, much like the spokes of a wheel, meaning that a large underground region can be accessed from a single surface well pad.

This means that with modern drilling techniques, far fewer well heads are needed on the surface to successfully tap a shale-gas field. This has made a big difference in the cost of accessing these so-called unconventional reserves, while also ensuring that the surface footprint of a drilling operation is far smaller. (I'll say more about surface impact a little later on.)

Once a well has been sunk and properly cased, the hydraulic fracturing process begins. A slurry of mostly sand and water is pumped down the well, along with a few chemical additives designed to keep the sand in suspension, improve lubrication, inhibit corrosion and dissolve scale. Under pressure, the water creates small cracks in the surrounding shale, which are then held open by the sand in the fluid. Once this process is complete, the gas trapped in the shale can be pumped to the surface. Wells may need to be hydraulically fractured, or 'stimulated', several times during their productive life.

Some of the fluid pumped down the well remains there, just as it would in deep-well injection used to dispose of hazardous waste. This stuff poses no risk. Once it's down there, it stays there. The idea that cracks from hydraulic fracturing might somehow reach up through several kilometres of rock, including layers containing brine aquifers, is preposterous.

Moreover, water, like everything else on the planet, is subject to gravity. Surface spills might pollute groundwater, but water doesn't flow uphill through miles of rock strata, even if they are permeable or full of cracks, to reach aquifers near the surface. In hydrology, university students learn about Darcy's law, which predicts, given the porosity of the soil or rock through which it travels, how fast groundwater will flow from a higher elevation to a lower elevation. It doesn't flow the other way around, and there's a minus sign in the equation to underscore this basic fact of physics.

That doesn't stop opponents of fracking from illustrating their propaganda with utterly misleading graphics. They usually omit brine aquifer regions, as well as the kinds of clay or rock with low or zero permeability, known as 'aquitards'. After all, how many casual readers who might get upset by industrial development know their aquitards from their aquicludes and aquifuges?

It is easy to design graphics that will mislead readers into believing that cracks formed by hydraulic fracturing in deep shale layers might reach freshwater aquifers. They often don't bother to show the multiple layers of casing that insulate wells from the surrounding aquifers either, in the hope that readers will conclude that well sleeves are easily breached. Compare Figure 1, which is not misleading, with Figure 3, an example of a propaganda graphic created by the Checks and Balances Project.

When the fracturing stage is completed, having created small cracks propped open by grains of sand, some of the pressurised fluid returns to the surface as 'produced water'. This needs to be treated with the same care as any other industrial by-product. Although most of the chemicals used in hydraulic fracturing are fairly inoffensive additives, such as those you might find in a swimming pool or in hand soap, others are not so benign. The produced water needs to be pumped or trucked away, either to water-treatment facilities for recycling, or to industrial-waste-disposal facilities designed to accept this kind of water.

Once the well has been drilled, it is capped. This is probably the riskiest part of the operation. When the drill is removed from the well to be replaced by a valve, pumping equipment and the connection to a gas pipeline, the pressure in the well can cause gas and fluids to escape from the well. Losing control of this process, known as a well blowout, is a costly disaster for drilling companies and a dangerous accident for the surrounding environment. Operators create a safety barrier around the well head to contain any fluid in case this does happen, but even though the damage from a blowout is typically localised, it can be fairly severe.

Surface spills – again, due to gravity – are the main cause of groundwater or stream pollution. Fairly effective clean-up is possible, but the process is costly and complex, so minimising accidents is a priority for drilling companies. Surface spills are a risk in any industrial or mining process, of course, and not only with hydraulic fracturing. However, that doesn't mean they're not serious when they do happen.

As with well sleeves, much technical work has been done to improve capping technology. Engineers even use the term 'green completion' to refer to a modern well-capping process that minimises gas escape and prevents well blowouts.

Once capped, the magic starts. Connect the well to a pipeline, and it can produce clean-burning, cost-effective natural gas for years, with little or no human intervention. Once the drilling rigs and fluid storage tanks have been removed, only a small well head – no more unsightly than a farm windmill – remains as a sign that here the earth's riches are being harvested to fuel human progress and prosperity.

Not everyone shares this optimistic vision of a productive future fuelled by natural gas, however.

2

Fracking

A Light on Scary Nightmares

FLAMMABLE WATER! THEY YELL

The iconic image of the dangers of fracking comes from a film billed as a documentary. Called *Gasland*, it is the second film by independent director Josh Fox. The scene involves a Colorado resident turning on a kitchen tap and then lighting the 'water' that comes out. It is spectacular film-making for sure, but what does it mean?

The film attributes the flames to methane, which is indeed a flammable gas. Methane is the reason why gas-drilling companies drill for gas. Further, the film blames the nearby hydraulic fracturing operations for the gas in the water.

This prompted the Colorado Oil & Gas Conservation Commission (COGCC) to issue a report to 'correct several errors in the film's portrayal of the Colorado incidents'.[1] It starts by describing methane, and making an important distinction to which we'll return at other points in this chapter:

Methane is a natural hydrocarbon gas that is flammable and explosive in certain concentrations. It is produced either by bacteria or by geologic processes involving heat and pressure. Biogenic methane is created by the decomposition of organic material through fermentation, as is commonly seen in wetlands, or by the chemical reduction of carbon dioxide. It is found in some shallow, water-bearing geologic formations, into which water wells are sometimes completed. Thermogenic methane is created by the thermal decomposition of buried organic material. It is found in rocks buried deeper within the earth and is produced by drilling an oil and gas well and hydraulically fracturing the rocks that contain the gas.

In Colorado, thermogenic methane is generally associated with oil and gas development, while biogenic methane is not.

The report proceeds to set out how it is possible to distinguish between these forms of methane based on their isotope signatures. Suffice it to say scientists don't dispute that it is possible to do so:

Gasland features three Weld County landowners, Mike Markham, Renee McClure, and Aimee Ellsworth, whose water wells were allegedly contaminated by oil and gas development. The COGCC investigated complaints from all three landowners in 2008 and 2009, and we issued written reports summarizing our findings on each. We concluded that Aimee Ellsworth's well contained a mixture of biogenic and thermogenic methane that was in part attributable to oil and gas development, and Mrs. Ellsworth and an operator reached a settlement in that case.

However, using the same investigative techniques, we concluded that Mike Markham's and Renee McClure's wells contained biogenic gas that was not related to oil and gas activity. Unfortunately, Gasland does not mention our McClure finding and dismisses our Markham finding out of hand.

To drive the point home, the report explains that Markham drilled his water well through no fewer than four coal beds, and notes that a report published in 1976 – the first of at least thirty such reports – warns that the aquifer from which he gets his water contains 'troublesome amounts of … methane'. Moreover, the nearby gas wells, the first of which was subjected to hydraulic fracturing in 1991, had never had any failure that could have caused seepage of fracking fluid or natural gas into the water aquifer. Nor did the McClure and Markham water contain any of the toxic chemicals – known as the BTEX complex, namely benzene, toluene, ethylbenzene and xylenes – usually associated with deep-shale gas. Twenty-seven other wells were tested in response to these complaints, and none, other than the Ellsworth well, was found to have contained any suspicious methane at all.

In a lengthy answer to his many critics, Fox admits that he doesn't know how the gas got into the drinking water, only that it did. Worse, at a screening at Northwestern University in Chicago, journalist and filmmaker Phelim McAleer asked Fox why he did not disclose the existence of the 1976 report. Clearly flustered, Fox says, 'Well, I don't care about a report from 1976. There are reports from 1936 that say people in New York could light their water on fire.'[2]

In short, Josh Fox admits that he simply ignored evidence that proves flammable water is nothing new in the area. The film footage was there for one reason only: as propaganda to scare viewers into believing that their water was likely to be contaminated.

While researching my first column on this subject, I thought I'd check Fox's claim that the only reason no link between hydraulic fracking and drinking-water pollution had ever been found was because it had 'never been investigated'.[3] I turned up dozens of reports from US states in which shale-gas drilling occurs and found many instances of investigations, but none of drinking-water contamination. I quoted a few examples in the *Daily Maverick*:

'After 25 years of investigating citizen complaints of contamination, [Ohio] geologists have not documented a single incident involving contamination of groundwater attributed to hydraulic fracturing,' swore John Husted, head of the Division of Mineral Resources Management of Ohio.

'While we do currently list approximately 421 groundwater contamination cases caused by pits and approximately an equal number caused by other contamination mechanisms, we have found no example of contamination of usable water where the cause was claimed to be hydraulic fracturing,' wrote Mark Fesmire of the New Mexico Energy, Minerals, and Natural Resources Department.

'After review of [our] complaint database and interviews with regional staff that investigate groundwater contamination related to oil and gas activities, no groundwater pollution or disruption of underground sources of drinking water has been attributed to hydraulic fracturing of deep gas formations,' testified Joseph J. Lee Jr, head of the Source Protection Section of the Division of Water Use Planning of the Pennsylvania Department of Environmental Protection.

'I can state with authority that there have been no documented cases of drinking water contamination caused by such hydraulic fracturing operations in our State,' concurs David E. Bolin, deputy director at the State Oil and Gas Board of Alabama.

'I sincerely hope that you [Scott Kell of the Ground Water Protection Council, in testifying before Congress] will clear up the misconception that there are "thousands" of contamination cases in Texas and other states resulting from hydraulic fracturing. The Railroad Commission of Texas is the chief regulatory agency over oil and gas activities in this state. Though hydraulic fracturing has been used for over 50 years in Texas, our records do not indicate a single documented contamination case associated with hydraulic fracturing,' wrote Victor G. Carrillo, chairman of that august body.

He is worth quoting further:'The Texas Groundwater Protection Committee [TGPC] tracks groundwater pollution in Texas. All Texas water protection agencies, including the Railroad Commission, are members. Each year, the TGPC publishes a Joint Groundwater Monitoring and Contamination Report. The 2007 report cites a total of 354 active groundwater cases attributed to oil and gas activity – this in a state with over 255 000 active oil and gas wells [11 000 of which use hydraulic fracturing]. The majority of these cases are associated with previous practices that are no longer allowed, or result from activity now prohibited by our existing regulations. A few cases were due to blowouts that primarily occur during drilling activity. Not one of these cases was caused by hydraulic fracturing activity.'

Colorado? Nada. Alaska? Nope. Kentucky, Louisiana, Colorado, South Dakota, Tennessee? Nothing.

According to Michigan's Office of Geological Survey (OGS): 'There is no indication that hydraulic fracturing has ever caused damage to groundwater or other resources in Michigan. In fact, the OGS has never received a complaint or allegation that hydraulic fracturing has impacted groundwater in any way.'

Same thing in Indiana and Oklahoma: no harm was ever found to have resulted from hydraulic fracturing, because they had never even received any reports of such a thing.

Not one instance. Not a single one, over more than half a century, covering hundreds of thousands of wells.[4]

Much is made in *Gasland* of the hundreds of dangerous chemicals used in hydraulic fracturing. Fox claims that almost 600 different chemicals are used. This is true, if you count all the chemicals and their variants that theoretically could be used. A hardware store stocks hundreds of different paints and varnishes, although they essentially reduce to half a dozen basic types, and you only need one for a given project. The same is true for gas drilling. In reality, only about a dozen additives are actually used per well.

Many opponents of fracking have claimed that these chemicals are secret. Granted, gas companies have historically been reluctant to give away proprietary information to competitors, but in response to public concern, a list of chemicals used in fracking, with names, descriptions, purpose and other common uses, was published by the US Department of Energy as early as April 2009.

Subsequently, an astonishingly detailed directory, specifying the exact chemical composition of fracturing fluids for any single well in the US, was published by the industry and made publicly available on a website at fracfocus.org.

This disclosure is the reason why headlines such as this one, from the *Colorado Independent*, are possible: 'Fracking chemicals found in Wyoming groundwater'. If true (and there is much cause to believe it is not), this would be anecdotal evidence which demonstrates that pollution caused by fracking could be proven in principle, despite Josh Fox's claims that nobody knows what chemicals drilling companies use.

Jonathan Deal, national coordinator of the TKAG, sent me that story as evidence that fracking had indeed polluted groundwater, contrary to the claims of industry, with which he assumes I concur. (I don't, of course. I've already noted surface spills and well blowouts as significant risks, which, though not caused by hydraulic fracturing nor unique to gas drilling in general, need to be prevented.) But a single case of fracking chemicals in the groundwater, even if true, is a far cry from proving that we have to choose between gas and water. It says nothing about the scale or frequency of pollution incidents, although it certainly seems troubling. Upon investigation, however, even this story turns out to be misleading.

The headline itself is, quite simply, false. It is based on a report by ProPublica, a public interest group none too friendly towards oil companies. ProPublica tells the story more accurately in its report title: 'EPA Finds Compound Used in Fracking in Wyoming Aquifer'. The difference is significant. First, only a single chemical is involved, not many. Second, the fact that this compound can be used in hydraulic fracturing does not mean gas drilling was the source. Identified as 2-butoxyethanol, this is a common solvent used in paints and varnishes, household cleaning products, fire-fighting foam, dry-cleaning chemicals, liquid soap, cosmetics, varnish remover, oil-spill dispersants, inks, ink removers and whiteboard cleaners. In fact, in many of these products it is the main ingredient. The Agency for Toxic Substances and Disease Registry at the US Centers for Disease Control and Prevention warns against skin contact or inhalation of large amounts of the stuff, but describes its environmental impact as, well, not very much:

> 2-butoxyethanol and 2-butoxyethanol acetate may be released into the air when they are used as solvents and in household products. In air, [they] may be removed by rain, ice, or snow. [They] may break down in air into other compounds within a few days. Both compounds may pass into air from water and soil. [They] do not build up in plants and animals.[5]

Given the high levels of other pollutants found in the region, the low risk posed by the one chemical that was found and that has many ordinary household and industrial uses, including in fracking, and the very curious fact that the samples contained not a single other fracking chemical, an initial assessment of this case suggests that the drilling company in question, the Canadian firm EnCana, is correct when it denies responsibility. In other words, the link to gas drilling is, given the evidence at the time of writing, utter nonsense. That Deal highlights such a trivial and inconclusive case as evidence to counter claims that gas drilling rarely if ever causes groundwater pollution demonstrates how hapless environmentalists clutch at straws in the hope of supporting their case.

Besides, how many environmentally minded people will bother to research 2-buto-whatever-it's-called? That it sounds scary is enough, even if the story plays fast and loose with the truth. The only purpose of such propaganda, as with Josh Fox's film, is to scare voters and legislators into calling for a ban on gas drilling.

When I thanked Deal for alerting me to this story and informed him that I would include it in this book as an example of the exaggeration of the anti-fracking lobby, his response spoke volumes. He proposed that I title this book *Memoirs of a Sycophant.* Evidently, facts and evidence don't matter when you're engaged in the heroic struggle against corporate tyranny, and any journalist who dares suggest otherwise is either in the pocket of the oil industry or an ignorant dupe.

In another apparent error, Fox attributed pollution in Pennsylvania, where fish died along a thirty-five-mile stretch of Dunkard Creek, to natural gas drilling. Unfortunately, the incident had been investigated, and the cause was found to be run-off from a nearby coal mine. The local press disputed Fox's version, noting that he'd never even visited the area. Fox responded that unnamed 'eyewitnesses' had told him fracking waste water had been pumped into the coal mines, and that is why the gas industry, rather than the coal industry, is guilty. The story prompted John Hanger, the secretary of the Pennsylvania Department of Environmental Protection, to describe *Gasland* as 'fundamentally dishonest' and 'a deliberately false presentation for dramatic effect' in an interview with the *Philadelphia Inquirer.*[6]

The same paper that disputed *Gasland's* version of the Dunkard Creek incident, the local *Observer-Reporter*, reported this: 'The threat of pollution created by the Marcellus Shale gas industry has resulted in the Monongahela River being named as one of the 10 most endangered rivers in the nation by a Washington, D.C.-based environmental group.' It takes the reporter seven paragraphs to get to the truth: 'The [Monongahela] River has been affected by the coal industry for many years. Further problems that might be created by the Marcellus Shale gas drilling could "send it over the top", said Emily Bloom of the Center for Coalfield Justice.'[7]

The upshot is that coal is dirty, and gas has, to date, proven not to be so. But just in case that changes, we're going to call the local river 'most endangered'. There's a word for this kind of thing. It's called propaganda. Some might even call it lies.

These aren't isolated examples. Elsewhere in the film, Fox warns that fracking in Wyoming is driving the local sage grouse to extinction. Odd, then, that the state's wildlife authorities are happy to permit hunting the endangered birds.

Gasland was nominated for an Oscar for best documentary, and won a special jury prize at the Sundance Film Festival. One surmises Hollywood types appreciate dramatic effect but have a hard time distinguishing fact from fiction.

The furore over well-water pollution prompted researchers at Duke University to settle the matter once and for all, scientifically. They selected sixty-eight water wells to test, and correlated their methane content with their proximity to gas-drilling sites. What they found was very little indeed.

'Scientific study links flammable drinking water to fracking,' crowed one newspaper after another.[8] Actually, no, it did not. Let's quote: 'Based on our data, we found no evidence for contamination of the shallow wells near active drilling sites from deep brines and/or fracturing fluids.'[9] How the phrase 'no evidence' in a study gets turned into 'links' in a newspaper headline is something editors must one day explain to me. Perhaps it has something to do with the political leanings of many journalists.

When a major South African news organisation, *News24*, selected a columnist to rebut my first column on fracking, they chose a fellow by the name of Andreas Späth to write it. He identifies himself at the end of his column as the holder of a degree in geochemistry and the manager of a small bookstore. No doubt he's a nice guy with good intentions and genuine fears about the dangers of hydraulic fracturing. However, what he fails to disclose in his biographical note is that he is the 'anti-fracking coordinator' for Earthlife Africa, a radical environmental lobby group. Surely this would be as relevant to the substance of his column as it would have been relevant if I had been in the pay of the oil industry?

Another series documenting the dangers of fracking was written by Geraldine Bennett for *Moneyweb*, a high-profile financial news site. At the end of her reporting, she disclosed that her extensive travels through the beautiful Karoo had been sponsored by local residents affected by gas drilling, and had in part been funded by the TKAG itself.

Jonathan Deal proposed to me that in order to attend a debate on gas drilling in which he was to participate, I solicit the necessary funding from Shell. Once again, the implication was clear: it is okay to work for, or be funded by, the environmental movement when debating an important issue of public interest, but if a journalist dares to do fact-checking, that journalist must be corrupt. Unlike the journalists who accepted money from Deal, I rejected his suggestion with the amused contempt that it deserved. It was gratifying to see a number of my readers jumping to my defence, letting Deal know that playing the man amounts to an admission that the ball is unplayable.

Now that we have a fairly good idea how those 'links' emerge in the popular press, let's return to the Duke University study that supposedly proved links between fracking and well-water pollution. The only fact of any significance that the Duke study found was that some water wells very near to gas-drilling sites contained significantly higher levels of thermogenic methane – the kind that comes from deep underground, instead of from shallow organic or coal sources.

It would have been easy to show that the gas came from failed or improperly cased wells if the water contained signs of telltale fracking chemicals. However, not a single well showed any trace of any of the chemicals typically used in hydraulic fracturing. Recall that these chemicals are disclosed in an online registry maintained by the gas-drilling industry, so the argument that pollution can't be proven because the chemicals are an industry secret is specious.

Because there were no baseline data prior to hydraulic fracturing, the study could not determine whether anything had changed, or whether there's just more gas in the ground where there's a lot of gas in the ground. After all, natural cracks and fissures that bring fossil fuels to the surface have historically been the signs that lead prospectors to sites that might be rich in reserves.

As there was no evidence of fracturing fluid in the water, the researchers were at a loss to explain how this gas they found might have got there. The study peters out feebly, with the conclusion that more research is needed. That's hardly the resounding proof of guaranteed groundwater pollution due to gas drilling that groups like the TKAG claim it to be.

But is the methane itself something to fret about? It turns out that it isn't. Gerrit van Tonder, a professor at the Institute for Groundwater Studies at the University of the Free State, told me that methane in the concentrations the Duke study found is very unlikely to be a health risk, other than by asphyxiation if it were dense enough. (The same is obviously true for the other gases that we breathe that aren't oxygen: pure nitrogen or carbon dioxide would cause death by asphyxiation. Neither is poisonous in drinking water. In fact, one of them we deliberately add to make drinks fizzy.)

The Duke report concedes that methane is not regulated as a drinking-water contaminant because it is not toxic, and only a few wells exhibited enough methane to require 'hazard mitigation'. Van Tonder explains that this is a common problem in water wells, and the simple solution is to install ventilated storage tanks in which the methane can flash off before consumption. The 'hazard' referred to in US regulations involves fire risk, not a health risk due to human consumption.

Given the many academics, journalists and environmentalists who are on the case to prove that hydraulic fracturing is something evil that will poison our drinking water, the paucity of anecdotal data, the distortions of the few anecdotes they *can* find and the complete lack of statistical evidence are, frankly, astonishing.

But wait, the critics cry. It's not the hydraulic fracturing itself, but gas drilling in general that causes pollution. This argument has merit. A little bit of merit. Like the last coins in your wallet when you thought you were broke.

REMEMBER THE RAVAGED NIGER DELTA, THEY WARN

Environmental activists like to cite Nigeria's oil delta as an example of why South Africa should not permit Shell to operate in this country. However, the comparison hardly holds up.

In Nigeria, Shell is a junior partner in a firm in which the majority stake is held by the state-owned Nigerian National Petroleum Corporation. The Nigerian government receives approximately 95 per cent of all after-cost earnings from the joint venture. This accounts for 80 per cent of the Nigerian government's revenue. By contrast, Nigeria accounts for less than 10 per cent of Shell's global revenue. This is not insignificant, but it seems reasonable to suggest that the bulk of the responsibility for the actions of the Nigerian joint venture must be laid at the Nigerian government's doorstep. After all, it has not only legislative, regulatory and enforcement power, but also effective management control over the oil operations in the delta.

Critics claim that the total amount of oil spilled in Nigeria is twice as much as that of the Deepwater Horizon oil-spill disaster in the Gulf of Mexico in 2010. This is, roughly speaking, correct. However, the spills in Nigeria took place over fifty years, a period which included the devastating Nigerian Civil War, better known as the Biafran War. Cumulative pollution over half a century is hardly comparable to a single event.

In the Niger Delta, Shell operates in a highly risky environment, in which staff have to be protected at all times from kidnap for ransom and pipelines are routinely sabotaged. Often, the damage is done by local black-market traders, who cause a spill and then go out in boats to scoop up the oil from the delta's polluted waters, to be turned into fuel at illegal refineries. More sophisticated thieves cause a spill, wait for the pipeline to be turned off for repairs, and dig underground to attach their own pipe to the main line.

Now one might easily imagine illegal whisky stills hidden in the forests, but entire crude-oil refineries? This sounds like a fantasy dreamed up by a Shell propaganda team. However, the Nigerian government routinely acts against illegal oil refineries, because it has financial interests of its own to protect. In March 2011, government forces found and destroyed over 500 such refineries in a single town.

The United Nations Environment Programme (UNEP), while often severely criticising Shell, has agreed in various reports that the majority of the spills can be attributed to circumstances beyond the company's control, including civil strife, deliberate sabotage, extortion for damages and illegal oil refineries.

Of course, none of this relieves Shell of reasonable obligations to minimise the harm it does where it operates. Indeed, the company takes significant risks with the lives of its workers by sending clean-up crews whenever spills are reported. Often they are kidnapped for ransom, or violently rebuffed by rebel groups who prefer to raise money from damages compensation than to see environmental pollution cleaned up. It seems only fair to consider these extraordinary circumstances when judging Shell's record in Nigeria.

None of these realities in Nigeria applies to South Africa. Although government corruption in South Africa is a significant risk to its people, there is no major political strife that is likely to spill over into violence or sabotage against companies that do business with the government. Abduction for ransom is a negligible risk in South Africa, environmental management and regulation of industrial and mining activity is fairly effective, and the justice system is robust. Moreover, natural gas is a very different product from crude oil, presenting a much-reduced risk of severe pollution. Comparing the two is comparing apples with oranges.

Most large companies at one time or another fall foul of the myriad of laws and regulations with which they have to comply. This can largely be attributed to the ever-growing complexity of the environmental, health, safety, labour and tax regime. Besides, despite the best intent in the world, no company can guarantee that none of its employees will act negligently, recklessly or with criminal intent, any more than any government can offer ironclad guarantees to its people that none of its officials will ever act incompetently or be corrupted, or that none of its citizens will, despite the law, commit theft, fraud, assault or murder.

Humans are fallible, no matter whether they act as private citizens, government employees, corporate workers or, indeed, environmental activists. Oversights, irresponsible behaviour and outright acts of malfeasance happen. When they do occur, we ought to respond by expecting prompt and full remediation, compensation for those who suffered damages, criminal action against transgressors, and the establishment of procedures and regulations to prevent a recurrence of the problem.

A good analogy is airline accidents. They happen, and when an accident takes place it is really unfair to the passengers on that aeroplane. However, horrific photos of crashes are not good grounds to ban flying. They represent anecdotal evidence of mishaps that occur rarely. That's not to say we shouldn't care about accidents and simply accept them as an unavoidable fact of life. When negligence, reckless behaviour, corruption or criminal oversight was responsible, it is reasonable to take tough action against the transgressors. And even if there is no blame to be attached to the airline in question, the rational response is to find out what went wrong and try to make sure that it never happens again. This we do. The risk of flying exists, but it is low and manageable.

Large companies of all kinds, but especially in high-profile industries such as petroleum extraction, employ entire floors full of lawyers, bureaucrats, accountants and experts on fields ranging from environmental management to occupational health and safety. This overhead cost is incurred to help companies meet – and often to exceed – the regulatory demands of the countries in which they operate. Moreover, there is a major financial incentive for companies to minimise the risk of civil claims and regulatory fines, as well as a very significant marketing incentive to avoid damaging publicity.

Operations in the developed world demonstrate that most multinationals are capable of conducting themselves with the highest integrity, provided that the local legal and regulatory framework is conducive to it. Even in the developing world, the data show that environmental quality improves as foreign investment increases. A startling example can be found in the much-maligned China, where, according to the UNEP, air quality as measured by particulate matter concentration was 50 per cent worse in 1990 than it is today, while foreign direct investment has risen from almost zero to $80 billion. The declining pollution trend in China is ongoing.

It is true that no company is perfect, and critics can always find something to complain about. However, this is not a good reason to prevent companies from producing the goods and services we need to grow more prosperous, and employing the people who need the jobs. The question is how to prevent accidents and transgressions, and how to deal with them when they do happen. Ultimately, the issue of corporate responsibility does not rely on whether or not you trust a company. Trust may be earned, but it shouldn't be presumed. As noted earlier, one should trust oil companies no more than one should trust the activists who agitate against them.

Ultimately, it comes down to the restraints imposed by a competitive market, to the very real publicity risk presented by an effective media, and to whether or not South Africa has the legal and regulatory institutions to hold both local companies and foreign firms operating in the country accountable. If those opposed to oil companies claim that South Africa does not, then the country has far more serious problems than the risk of pollution in a remote, arid region.

For example, consider the production and transport of milk. Milk is 400 times worse as a pollutant than sewerage. When it gets into rivers and streams, it is pretty vicious, killing most fish life as the bacteria that break it down suck the oxygen out of the water.

Dairy farms and milk-bottling plants work under strict regulations to ensure that the waste water from the operation does not pollute surface water and underground aquifers. Operators are subject to large fines, and even to having their right to operate taken away, if they contravene the rules on how to handle the transport of milk and the disposal of production waste.

Still, accidents do happen, sometimes as a result of deliberate neglect on the part of greedy farmers. But because spills happen, do we ban the production of dairy products?

With regard to claims of pollution caused by hydraulic fracturing, almost all the anecdotal stories involve single accidents where chemical storage or waste-water disposal operations went awry. These stories involve surface spills, ranging from well blowouts in the worst case to transport or storage-related accidents. But although such accidents do occasionally happen, when case studies are cited, the same few isolated instances always make an appearance. Yet handling dangerous chemicals and industrial waste is not a new problem. It is a well-understood process, subject to perfectly ordinary rules and regulations to prevent them from harming the environment or the health of people living near the operation.

Waste water from shale-gas drilling operations may need special treatment, but why would this be an insurmountable burden? Government has every right to enforce proper waste-water treatment, and routinely does so in the case of farms, factories and mining operations. But, the opponents of fracking say, we can't trust energy companies to do the right thing, can we?

Well, can we?

Near Cape Town stands a large oil refinery, operated by a big oil multinational, Chevron. As part of its routine operation, this refinery produces large amounts of waste water. After being treated according to the required environmental standards, the waste is discharged into the sea from a pipe that runs some 500 metres offshore. Nobody on land notices, the water does not harm marine life, and the company deals with the waste in full compliance with all environmental regulations.

Enter kitesurfers, who actually do end up 500 metres offshore in Table Bay, without a boat under them. The waste from Chevron's pipe had an odd colour and smell, which they did not like. They complained.

If the reputation of oil companies was anything to go by, Chevron would have laughed the surfers off. The company has all the compliance certificates it needs, and there is no evidence of any harm being done to anyone. But Chevron chose to build the largest waste-water treatment plant in South Africa, at a cost of R107 million. It didn't have to. But it did it anyway, to alleviate 'the nuisance impact that the wastewater was having on kite surfers'.[10]

Merely observing that fracking chemicals are harmful, and that waste water is mildly toxic, does not make the case that shale-gas drilling should be banned. And the argument that we can't trust oil companies seems like a serious case of prior restraint – prohibiting something that could conceivably pose a risk, rather than acting prudently to avoid accidents and respond appropriately on the rare occasions when they do happen.

TANKERS FULL OF RADIOACTIVE WASTE! THEY PANIC

Another fear is that the water produced after fracking might contain radioactive substances. It does, but not in alarming amounts. Radio-active substances are a reality for any mining activity, because of naturally occurring radioactive material in rock. This is why you're more exposed to radiation in a cave than above ground, and why Johannesburg's iconic gold mine dumps weren't ideal locations to set up residence.

Radioactivity comes in many varieties. All of it involves the 'decay' of the nucleus of an atom. Usually, this atom is an unstable variation, known as an isotope, of a normally stable element.

This process gives off high-energy photons (gamma radiation), electrons (beta radiation) or entire helium particles (alpha radia-tion). Some of these are easily blocked but are fairly massive, while others penetrate deep into materials, causing far more subtle damage at a molecular level.

The longevity of radioactivity is measured in half-lives. As a substance decays, the original substance gradually grows less, while the resulting by-products of radioactive decay increase. Since each decay event is random, scientists measure the time after which half of the original substance is left. So, if a substance has a half-life of a week, only a quarter of it will be left after two weeks, and one-sixteenth after four weeks. After a few months, almost nothing will be left of the original element, although in some cases the by-products of decay are themselves radioactive.

For example, ordinary carbon, known as carbon-12, has six protons and six neutrons. An uncommon isotope, carbon-13, has an extra neutron, but is also stable. A very rare isotope, carbon-14, has eight neutrons in its nucleus, is radioactive and has a half-life of 5730 years. One of its neutrons decays to form a proton and an electron. The electron is emitted as beta radiation, while the proton remains behind, changing the base element from carbon-14 to ordinary, harmless nitrogen-14. Carbon-14 is formed in the atmosphere by cosmic radiation and is absorbed in trace quantities by living tissue, but decays after death until, after millions of years, there is almost none left in deep rock. Because of this life cycle, carbon-14 is the basis for so-called carbon dating, invented in 1949.

Half-life gives an indication of how long a radioactive substance hangs around, but it says little about the intensity of the radiation it gives off. Radiation intensity is measured in various ways. One of them is a unit known as the becquerel (Bq), which measures the number of decay events caused by a substance per second. It is an international standard by which the intensity of radiation from different amounts of differing substances can easily be compared. The human body gives off about 4400 Bq from naturally occurring radioactive substances.

The radioactive risk from hydraulic fracturing is described as follows by the Groundwater Protection Council, as commissioned by the US Department of Energy and quoted by the UK House of Commons Energy and Climate Change Committee:

Some soils and geologic formations contain low levels of naturally occurring radioactive material (NORM). When NORM is brought to the surface during shale gas drilling and production operations, it remains in the rock pieces of the drill cuttings, remains in solution with produced water, or, under certain conditions, precipitates out in scales or sludges. The radiation from this NORM is weak and cannot penetrate dense materials such as the steel used in pipes and tanks.

Because the general public does not come into contact with gas field equipment for extended periods, there is very little exposure risk from gas field NORM. To protect gas field workers, [the Occupational Health and Safety Act] requires employers to evaluate radiation hazards, post caution signs and provide personal protection equipment when radiation doses could exceed regulatory standards. Although regulations vary by state, in general, if NORM concentrations are less than regulatory standards, operators are allowed to dispose of the material by methods approved for standard gas field waste.

Conversely, if NORM concentrations are above regulatory limits, the material must be disposed of at a licensed facility. These regulations, standards, and practices ensure that shale gas operations present negligible risk to the general public and to workers with respect to potential NORM exposure.[11]

Such naturally occurring radioactive materials are common in many environments, and even in our foods. In the case of mining, the most common risk is radium-226, which has a half-life of 1 602 years and is the basis for one of the oldest measures of radioactivity, the Curie. This substance was once commonly used in luminescent watches and instrument dials.

Many foodstuffs contain radium-226, potassium-40 or their decay elements. Bananas, for example, are a natural source of potassium, including of the radioactive variety, which has a half-life of 1.25 billion years. Brazil nuts contain both potassium-40 and radium-226, and are three times as radioactive as bananas. These foods are perfectly harmless, of course, even though the radiation is strong enough to set off detectors at airports and seaports, and the stuff accumulates in human bones, making us slightly radioactive ourselves. Table 2.1 gives a taste (if you'll excuse the terrible pun) of some typical radioactive sources, for comparison.

Table 2.1
Sources of radioactivity

1 adult human (65 Bq/kg)	4 500 Bq
1 kg of coffee	1 000 Bq
1 kg superphosphate fertiliser	5 000 Bq
1 kg of bananas, carrots or potatoes	125 Bq
The air in many 100 m² European homes (radon)	up to 30 000 Bq
1 household smoke detector (with americium)	30 000 Bq
Radioisotope for medical diagnosis	70 million Bq
Radioisotope source for medical therapy	100 000 000 million Bq (100 TBq)
1 kg 50-year-old vitrified high-level nuclear waste	10 000 000 million Bq (10 TBq)
1 luminous EXIT sign (1970s)	1 000 000 million Bq (1 TBq)
1 kg uranium	25 million Bq
1 kg uranium ore (Canadian, 15%)	25 million Bq
1 kg uranium ore (Australian, 0.3%)	500 000 Bq
1 kg low-level radioactive waste	1 million Bq
1 kg of coal ash	2 000 Bq

Source: World Nuclear Association, Idaho State University.

As a share of our total exposure to radiation, natural sources account for as much as 85 per cent. Most of the rest of a typical human's lifetime exposure to radiation comes from medical scans. Industrial waste and nuclear power present almost zero exposure risk to the general public, whether via drinking water or any other medium.

One must, of course, concede one fact to the environmental lobby: 'Radioactive Water!' makes a superb slogan for a poster at an anti-fracking demonstration. Only a real amateur would pass up such a splendid opportunity to play on the unfounded fears of the public. And still, the scaremongering is not at an end.

3

Fracking
Crying Wolf

THE KAROO IS BONE DRY, THEY WARN

Lewis Pugh put it plainly in his speech, when he said that the fracking debate is about which is more important: water or gas. He was talking not only about pollution, but also about the amount of water that the fracking process requires.

However, Pugh's dilemma is not really the choice before us. The choice is whether we ought to permit companies to use some water in order to drill for gas.

How much water? That's a moving target, but let's try to get some perspective.

Maarten J. de Wit of the Nelson Mandela Metropolitan University reports that a horizontal, multi-directional well would use between 10 million and 20 million litres of water. The Groundwater Protection Council in the US told its Department of Energy the number was between 2 and 4 million gallons, or 7.5 million to 15 million litres. Let's settle for 20 million litres, lest we be accused of exaggerating how low the impact of gas drilling will really be. This is also the number that the TKAG uses, not wishing to make things look less serious by acknowledging the lower end of the range.

Twenty million litres sure sounds like a lot, but how much is a lot, really?

The Vaal Dam, says De Wit, contains 25 trillion litres. A little arithmetic shows that you could frack over 6 000 wells with half a per cent of the Vaal Dam's water.

A golf course, according to a firm that builds irrigation systems, goes through 20 million litres of water in about ten days. From Aberdeen to Zwartkops, there are 430 golf clubs in South Africa. Every year, these golf courses consume enough water to frack around 15 000 wells.

Estimates of the share of a region's water supply consumed by hydraulic fracturing operations differ. The aforementioned Groundwater Protection Council says between 0.1 per cent and 0.8 per cent. Elsewhere, I've seen figures of up to 1.7 per cent in a dry region such as Texas.

This is certainly not a trivial amount, but is it really catastrophic? By comparison, agriculture consumes 75 per cent of South Africa's water supply. One wouldn't want to ban farming, of course, but complaining about the 'millions of litres' used for gas drilling turns out to be exaggerated hype when seen in perspective.

In its environmental management plan, Shell has committed not to compete with Karoo residents for water. The company has floated alternative options, such as using seawater, trucking in water from less-arid regions and using the Karoo's own deep brack-water aquifers. In addition, gas-drilling companies are continually improving on ways to minimise water use and to recycle waste water whenever possible. As De Wit points out:

> Karoo farmers and municipalities are right to be extra alert over their water rights. However, fracking technology has also moved on over the last [two to three] years; less water is now needed

and today water with a salinity of [twice that of seawater] can be used … In some instances, water is now dispensed with in the later stages of fracking and the gas is used to continue the fracking. Many technologies and best practices that can minimise the risks associated with shale gas development are already being used by some companies, and more are being developed.[1]

In a debate with Jonathan Deal of the TKAG, the chairman of Shell South Africa, Bonang Mohale, told an audience that the company would use 'brackish water 2000 metres below ground' and 'will recycle up to 50 per cent of the water'. He added the startling claim that '[a]t the end of the day there will be a net gain of water in communities'.[2] This is a claim that is worthy of scepticism, but it does suggest that the opposite extreme – water shortages leading to actual conflict – is a ridiculous notion.

The truth is that the claims by the green lobby are based on some superficial grains of truth, but they mask wild exaggeration about the scale of the threat to South Africa's drinking-water supply. South Africans won't be going to war because they sacrificed all their water for gas.

GAS IS WORSE THAN COAL, THEY CLAIM

One of the great advantages of natural gas over oil or coal is that it burns much more cleanly. It produces half as much plant food (carbon dioxide, which many climate scientists blame for global warming) as coal. This is the number that proponents of the fuel, not wishing to exaggerate, usually use.

However, it would not be an exaggeration to note that burning natural gas instead of coal reduces real pollutants, such as carbon monoxide and nitrous oxides, by a whopping 80 per cent, and that it almost completely eliminates pollution by sulphur dioxide and particulates. Table 3.1 lists substances emitted by burning different kinds of fuel, showing that gas emits half as much carbon dioxide as coal, and slashes the output of most pollutants even more.

Table 3.1
Combustion emissions by source
(in kg/billion BTU energy input)

Substance	Coal	Oil	Gas
Carbon dioxide	458 562	361 558	257 941
Carbon monoxide	459	73	88
Nitrogen oxides	1 008	988	203
Sulphur dioxide	5 712	2 474	1.3
Particulates	6 049	185	15
Formaldehyde	0.49	0.49	1.65
Mercury	0.035	0.015	0

Source: Energy Information Administration, 1998.

This positive picture of natural gas poses a grave problem for the environmental lobby. It has long pinned its hopes on the 'peak oil' theory of catastrophic resource depletion in order to advocate an end to reliance on petroleum, and the dawn of a new age of expensive but 'renewable' energy alternatives such as wind, wave and solar power. 'Peak oil' refers to the unknown point in the near future at which some economists and many environmentalists have warned that petroleum production will peak, before scarcity begins to reduce the industry's output and oil becomes gradually more expensive until, eventually, it makes pedal power look cost-effective.

It was hard enough to wait for the oil market to collapse suddenly, as the green lobby expected would happen when producers 'ran out' of oil, when along comes this new fossil fuel that is not only spectacularly abundant all over the world but is also much cleaner than any other. Politicians and other advocates of its use – including *Scientific American*, a well-respected publication among environmentalists – were positioning it as a 'bridge fuel'. The idea was that while the world waits the many decades it will take for technology to develop sufficiently so that some zero-emission sources of energy might be able to compete on both price and volume with traditional fuels such as coal and oil, natural gas offers the promise of a much cleaner transition fuel for electricity generation, heating and even some transport.

Then along came a professor from Cornell University, who rushed into print a study that claims natural gas is actually dirtier than coal. The premise is that methane is a fairly potent greenhouse gas (which it is), and therefore contributes significantly to the threat of climate change. If you include all the 'fugitive' emissions of methane during well drilling, capping, transport, transmission and processing, the Cornell study argues, the process of using natural gas is worse for global warming than the extraction and use of coal.

In a UK House of Commons report, the Geological Society floated the opinion that fugitive methane emissions during shale-gas exploration and production are 'very unlikely to be due to hydraulic fracturing, since this occurs at depths of several thousand metres beneath the surface'.[3] Not so, says the Cornell study.

Let's for the sake of argument assume that climate change is a crisis that is caused by humans and requires the world to immediately and drastically reduce greenhouse gas emissions. (We'll cover that issue in more detail in a later chapter.) If so, the Cornell study would appear to be very damaging. The green lobby duly latched on to it, in much the same way that it waved the Duke groundwater contamination study about as evidence against gas drilling. The lead researcher, ecologist Robert Howarth, told his university's newspaper:

> The take-home message of our study is that if you do an integration of 20 years following the development of the gas, shale gas is worse than conventional gas and is, in fact, worse than coal and worse than oil. We are not advocating for more coal or oil, but rather to move to a truly green, renewable future as quickly as possible. We need to look at the true environmental consequences of shale gas.[4]

This quote is revealing, because it demonstrates unambiguously that Howarth considers himself to be in the business of advocacy, not mere research, and that this advocacy means rejecting all fossil fuels in favour of renewables. Even the US Environmental Protection Agency weighed in: 'This study ... is an important piece of information that we need to bring into the discussion,' its deputy administrator, Robert Perciasepe, reportedly told a Senate Environment and Public Works Committee hearing on gas drilling.[5] Closer inspection of the study, however, shows that it is as flimsy as it sounds.

Howarth has a bit of a reputation. In a 2010 study along similar lines, he and his co-author, Tony Ingraffea, didn't realise that coal mining also produced methane emissions. In an energy sector that includes 'coal-bed methane' among unconventional gas resources, such an oversight in a study that purports to scientifically compare the emissions of coal and natural gas seems quite serious. The Cornell team promptly withdrew the study, and although many documents refer to it, the paper can no longer be found on Cornell's website.

But let bygones be bygones. Having been wrong in the past doesn't mean Howarth and Ingraffea are wrong now.

Or does it? Much was made of the fact that although the media were rushed prepublication copies of the study, it was published in a journal that is peer-reviewed. Why, then, did peer reviewers not spot the most glaring error in the study? Lay readers of the mainstream media, which splashed the study all over the headlines, might not be able to distinguish between gigajoules and kilowatts, but Michael Levi, the Senior Fellow for Energy and the Environment at the Council on Foreign Relations, a US think tank, can and did.

Here's why the distinction matters. Howarth compared the emissions for producing a given amount of gas, relative to producing the same amount of coal, in terms of their energy content as measured in gigajoules. However, the key question is how emissions compare per unit of electricity generated, which is measured in kilowatt-hours. 'This is an unforgivable methodological flaw; correcting for it strongly tilts Howarth's calculations back toward gas, even if you accept everything else he says,' writes Levi.[6]

There are many other problems with the study. Howarth and Ingraffea changed the accepted standards for comparing greenhouse-gas impact over time so that natural gas is weighted 45 per cent more heavily than even the UN Intergovernmental Panel on Climate Change (IPCC) does. This seems to be a completely arbitrary decision designed solely to reach the conclusions the researchers desired. The study bases its estimates for losses into the atmosphere on industry reports about 'lost and unaccounted for gas', or 'LUG', but fails to recognise that almost two-thirds of this gas is used to power its own transport, according to a lecture given by Timothy Skone, a US Department of Energy engineer, at Cornell University itself.

Nor did the study take into account that much of the rest of the gas is flared off and not simply vented into the atmosphere as pure methane. 'Howarth et al. ignore reports that flaring is common and assume, without providing support for their assumption or even stating it, that this gas is never flared from shale gas wells,' writes David McCabe, an atmospheric scientist. 'We're not fans of flaring, which still pollutes, but it is better than venting and reduces methane emissions a lot.'[7]

More than a dozen major studies have appeared that demolish the work of Howarth and Ingraffea. Among them, a recent study from the University of Maryland found the same errors mentioned above. It reiterates that the greenhouse-gas potential of natural gas used for electricity generation is only 56 per cent of that of coal, and adds: '[A]rguments that shale gas is more polluting than coal are largely unjustified.'[8] Another study, by researchers from Carnegie Mellon University, found the same. One of its authors, Paula Jaramillo, told the *Politico Pro* online magazine: 'We don't think [Howarth and Ingraffea are] using credible data and some of the assumptions they're making are biased. And the comparison they make at the end, my biggest problem, is wrong.'[9] A study by Wood Mackenzie, a Scottish firm that does research for the energy, metals and mining industries, reached a similar conclusion: 'Our analysis indicates that the Cornell study overestimated the average volume of natural gas vented during the completion and flowback stages by 60–65%. We conclude that the Cornell study over-estimated the impact of emissions during well completions by up to 90%.'[10]

You might expect industry-aligned groups to be piling it on in a fit of partisan politicking, but they're not alone. The atmospheric scientist quoted earlier, David McCabe, works for the Clean Air Task Force, which is a non-profit organisation dedicated to, well, clean air. It opposes all fossil-fuel use, including natural gas. He was scathing in his condemnation of the Cornell study: 'This paper is selective in its use of some very questionable data and too readily ignores or dismisses available data that would change its conclusions.'[11]

If even sworn opponents of natural gas don't believe the Cornell study, why should anyone else?

But the most revealing evidence comes from Howarth and Ingraffea themselves. According to John Hanger, former secretary of the Pennsylvania Department of Environmental Protection, they addressed a presentation to colleagues on 15 March 2011, in which Howarth called the data used in his study 'lousy', 'really low quality', 'teased apart out of PowerPoint presentations here and there'. Ingraffea said, 'We are basing this study on in some cases questionable data.'[12]

Writes Hanger: 'In a court of law, those would be case determining admissions. In a boxing match, the ref would stop the match. In science, they are an expressway to junk.'

Ingraffea told his colleagues: 'I hope you don't gather from this presentation that we think we're right.'

No, I think we can agree. It's junk. Prize junk, written by biased scientists who set out to reach predetermined conclusions and were willing to make any assumptions necessary to reach them – and blame the industry for 'bad-quality data' if they get called to account. Pity that the headlines in *TIME* magazine, the *New York Times* and a hundred other newspapers will survive forever as bullet points in the anti-fracking lobby's presentation slides.

In a public debate on fracking, hosted by Rhodes University in Grahamstown on 2 July 2011, I explained all of this to Jonathan Deal. Recently he sent me a message to visit the TKAG website if I wanted to know the facts about fracking. I did so. Under 'fracking facts', I found this: 'Shale gas may be cleaner burning than for example coal or oil, but if the life cycle emissions are compared to conventional fossil fuels (such as oil and coal) natural gas may be equally harmful to the atmosphere. This was confirmed in a recent study by Cornell University in America.'

That, I'm afraid, is not a 'fact'. It is wilful propaganda that cannot be attributed to ignorance, but can only be explained as a deliberate lie designed to sway public opinion and influence public policy. That this lobby group, which ostensibly acts in the public interest, has any credibility left beggars belief.

THEY'LL DESTROY THE KAROO! THEY SCREAM

A major element in the opposition to gas drilling is the notion that the impact on the surface environment of the Karoo will be devastating. The exploration licence applications cover a vast area, which one day may contain thousands of gas wells. All of these wells will have to be drilled, fracked and maintained, requiring access roads, machinery, waste-water infrastructure and lots of heavy trucks.

In a 2011 interview on a television channel in South Africa, the chief anti-fracking lobbyist, Jonathan Deal, had a large laminated poster under his arm. This poster, shown in Figure 4 (the added title is mine, not Deal's), is often used to illustrate the visual surface impact that gas drilling by means of hydraulic fracturing could have on a region. The impressive aerial photograph was taken by Bruce Gordon of EcoFlight for a group known as SkyTruth, which provides the useful service of producing such images to draw attention to the surface impact of various industrial and environmental activities, including gas drilling.

As with most of the other 'evidence' we've examined from the anti-fracking lobby, there are a few issues with this photograph that make it deceptive as a representation of the impact of shale-gas drilling with hydraulic fracturing.

The photograph was taken on 12 May 2006, and represents a particularly productive region known as the Jonah gas field in the

Green River Basin in south-western Wyoming. The Jonah field covers eighty-five square kilometres, which is the size of a small town. It is remarkable for containing as much as 10 trillion cubic feet (tcf) of gas, despite its relatively small extent. It is operated by EnCana Corporation of Canada and BP of Britain, along with a few smaller local partners.

If you've been paying attention to the fracking debate, you'll notice that the photograph dates from well before the major unconventional gas plays, like the Marcellus Shale in Pennsylvania and the Barnett Shale in Texas, began to boom in the US between 2007 and 2009. This is relevant because this photograph does not, in fact, depict shale-gas drilling using hydraulic fracturing and horizontal drilling, as is being proposed in the Karoo. The Jonah field is a tight-sands operation and employs mostly vertical wells, which significantly increases the well-head density on the surface.

Aside from the fact that this kind of density will not be used in the Karoo, the notion that energy companies will blanket the entire prospecting area with such fields is nonsense. They'll seek out particular 'hot spots' and deploy wells only where the shale is most productive. In the vastness of the Karoo, the wells will be nigh on invisible.

It is true, however, that there will be some surface impact. That stands to reason. But how bad will it get? If you believe the TKAG, the Karoo will be ruined by thousands of trucks, access roads, heavy machinery and unsightly wells. Is there a way to find out what a worst-case scenario could really look like?

As it turns out, there is. Among the concerns is whether or not gas drilling will interfere with South Africa's bid for the Square Kilometre Array (SKA). This is a radio telescope installation project that will be awarded to one of two bidding countries – Australia or South Africa – and will involve a capital budget of €2 billion. (At the time of writing, the decision had not yet been made.) The aim is to set up a network of three different types of radio telescope with a total collection area of one square kilometre, making the array the world's most sensitive telescope by far.

A massive region in the Karoo, comprising much of the Northern Cape province, has been set aside as an 'astronomy reserve' (see Figure 5), and is protected by law from disturbances that could interfere with optical or radio astronomy, ranging from aeroplanes and heavy trucks to arc welding. A demonstration site, with seven telescopes, has been built near Carnavon in the central Karoo as part of the bid process for the larger project. It is known as the KAT-7 interferometer array, or the 'precursor' array, and it will ultimately form part of the sixty-four-dish MeerKAT installation, whether or not South Africa wins the bid to host the full-scale SKA. MeerKAT itself will be one of the most powerful telescopes in the world, even if it doesn't serve as a precursor to the full SKA.

Adrian Tiplady, the SKA site characterisation manager, described the project to me during an interview for an article in *ITWeb Brainstorm*, a popular South African technology magazine:

> The core area of the SKA, approximately 80 km north-west of the town of Carnarvon, will have the highest density of receivers: approximately 1 500 dishes, together with a mixture of other hybrid receiver technologies, will be concentrated in a 5 km diameter core. This represents about 50 per cent of the total collecting area for the SKA. The remaining collecting area will be distributed in a spiral configuration [five spiral arms around the core], with clumps of receivers [stations] at ever-increasing intervals out to 3 000 km from the core. However, most [about 80 per cent] of the collecting area will be within 180 km of the core. A number of SKA stations, totalling approximately 10 per

cent of the collecting area, lie within the proposed western precinct [of Shell's exploration application].[13]

While Tiplady was apprehensive about the impact of the gas-drilling operations on the sensitive radio receivers, he seemed happy to accept Shell's assurances that it will comply with all aspects of the law that protects the region's radio silence:

> Shale-gas drilling near the site introduces a risk. However, by going through the legal processes, which includes the Astronomy Geographic Advantage Act, we have an opportunity to strengthen the SKA bid by giving a case study of how we can protect the SKA through the AGA Act. If we didn't have the Act, and the cooperation of the various stakeholders, I would say it could have a potentially damaging risk.

The attentive reader will have noticed Tiplady's description of the SKA, however: thousands of radio telescopes, radiating out from a dense core with a radius of 180 kilometres to as far as Ghana, Kenya and Madagascar. Across this vast region will be scattered 3 000 radio telescopes of varying designs, in clusters of half a dozen or so each.

During the public debate on fracking held at Rhodes University, Professor Peter Rose, one of the participants, took the opportunity to explain to me that the SKA will, as the name implies, occupy only a single square kilometre, whereas gas drilling will happen all over the Karoo. Deal, too, accused me of ignorance. They appear to labour under the mistaken impression that the 'square kilometre' in the name of the installation refers to its geographic extent. In fact, the name refers only to the combined collection area of the instruments. In respect of the astronomy reserve, where

Tiplady said 80 per cent of the radio telescopes would be built, the error involves five orders of magnitude: the SKA will cover over 100 000 km². That, as some elementary arithmetic will confirm, is a little more than 1 km². In fact, it is more than 90 000 km², which is the total size of the three areas that Shell intends to explore in the hope of finding suitable drilling 'sweet spots'.

Each radio telescope cluster will occupy a site twice as big as a gas-drilling well pad. Unlike gas wells, these sites will be paved, but like gas wells, they will require access roads and support buildings. Whereas gas wells need pipes to tap off the natural gas, radio telescopes need electricity and a massive high-speed data backbone to connect the network together. Construction times for gas drills and radio telescopes will likely be comparable: on the order of a few weeks each, after which they both continue to operate with minimal maintenance activity. But while gas drillers don't impose major restrictions on nearby residents or traffic, the radio telescopes need to be isolated, dark and free of any radio interference.

The photographs of the construction of the KAT-7 precursor array in Figures 6 to 9 give an idea of the scale of the SKA project. Note the huge trucks, the substantial size of the telescopes, the appearance of a single cluster of telescopes and the scale of the core area on a map.

Based on this data and these images, I took the liberty of drawing an unauthorised artist's impression of what the core area of the SKA will really look like, including access roads, tarred radio telescope pads and support buildings (see Figure 10). Like the Jonah gas field, the area pictured is about 85 km², or roughly as big as a medium-sized Karoo town. Like a typical shale-gas field, it will be vanishingly small in the vastness of the Karoo. How many Karoo towns can you spot on a satellite map of South Africa?

The area depicted in the astronomy reserve map in Figure 5 appears to be empty, but it contains the towns of Loxton, Fraserburg, Carnarvon, Williston, Brandvlei, Vanwyksvlei, Halfweg and Kenhardt. No doubt these towns, although invisible from space, are significant to their residents and surrounding farmers. Surely the same importance ought to be attached to this region as the TKAG attaches to the supposedly fragile ecosystem that the gas-drilling companies are after?

Why, then, are environmentalists and local residents not protesting furiously about the construction of this massive telescope project? Since the surface impact is not much different, the only possible reason for the difference in approach is that heavy engineering for the sake of science is emotionally more attractive than heavy engineering for the sake of energy.

There are good reasons to support the SKA project, but also good reasons to oppose it. Among these are that the project involves a great deal of taxpayer money without employing a great number of people, and the fact that it requires strict regulations to prevent radio interference, which will severely restrict the activities of farmers in the region. The damaging impact it will have on the Karoo surface environment, however, is not a good reason to object. It would be nice if environmentalists were consistent, and withdrew their objection to gas drilling on the grounds of surface impact in the Karoo.

GAS WILL RUN OUT TOO, AND IT'S UNPROVEN, THEY SAY

A stock response by environmentalists in the fracking debate is to point to the fact that all fossil fuels, including gas, are in principle finite, and that the potential extent of the Karoo resources is unproven, so that potential benefits cannot be accurately weighed against costs.

This is probably the silliest of all the anti-fracking arguments. Of course the gas reserves are unproven! That's why the gas-drilling companies are applying for exploration permits, so that they can find out how much, exactly, is there.

Estimates by the US Energy Information Administration put South Africa's technically recoverable gas resources at 485 tcf (see Table 3.2). This estimate may well prove to be optimistic but, conversely, this is about a quarter of the total estimated to actually be underground, so future technology improvements may raise this number.

Table 3.2
Technically recoverable shale-gas resources (tcf)

1	China	1 275
2	United States	862
3	Argentina	774
4	Mexico	681
5	South Africa	485
6	Australia	396
7	Canada	388
8	Libya	290
9	Algeria	231
10	Brazil	226
11	Poland	187

Source: Energy Information Administration, 2011.

One tcf can generate 100 billion kilowatt-hours of electricity. This is equivalent to the maximum expected output of the giant new Medupi coal-fired power station for two and a half years. It is enough to provide every single household in South Africa with a free basic electricity grant of fifty units a month for sixty years. It could fuel 12 million gas-driven vehicles for a year. In short, a tcf is a lot of gas. In fact, one tcf of recoverable reserves was sufficient for the Mossgas project near Mossel Bay to get the go-ahead.

South Africa currently consumes about 0.19 tcf of natural gas per year, of which it produces 0.07 tcf and imports the rest. Assuming that the country builds a large number of gas-fired power stations to take advantage of the riches beneath its soil, it will have enough gas for a century or two.

It is true that not all promising fields turn out to be as rich as initial estimates suggest. This appears to have happened in the Marcellus Shale in the north-eastern United States, for example. Although the estimates of ten years ago said the vast region would yield only a paltry 2 tcf of technically recoverable gas, more recent estimates put it at a 410 tcf bonanza. This has since been scaled back dramatically, to 84 tcf. If this reduced estimate is accurate, it would represent a great blow to energy investors, although the region remains a rich reserve.

There's another possible explanation for the apparent over-statement of resources in the Marcellus Shale. By letting outsiders believe they expect to produce more gas than they think is actually there, gas-drilling companies discourage competition from other sources of energy. Imagine being a nuclear energy investor, faced not only with the public backlash against nuclear power after the disaster in Fukushima, Japan – we'll hear more about that in a later chapter – but also with big talk about cheap and abundant natural gas plays. Wouldn't you think twice about risking your capital against such formidable competition? Half the game is convincing your competitors you're stronger than you really are. Gas companies benefit by keeping their rivals timid, rather than encouraging aggressive competition. Meanwhile, their own investment decisions may well be informed by qualifications and doubts that they wisely keep secret. If investors in a firm that operates this way aren't in on it, they may have grounds to claim they were deceived, but otherwise the company's estimates are not really anyone else's business.

Whatever the truth, wild speculation either way about South Africa's gas resources is entirely immaterial until actual exploration has begun. No energy companies are committing to any predictions of how much gas there might be under the Karoo, or whether it will be worth extracting. However, even if the results of exploration indicate that South Africa's recoverable gas reserves are much lower than initially expected, they may still represent a significant prospect that could prove to be economically viable. To find out, we have to permit exploration.

Like the true extent of the reserves, whether or not they are economically viable is not really anyone else's concern. The gas-drilling companies will do the numbers and their investors will take the risk. Any losses will be borne not by South African consumers or taxpayers, but by the shareholders in the energy companies. It is their responsibility to check industry estimates, and to hold the companies they own accountable.

Why should an outside environmental lobby like the TKAG – with its prominent warnings of a 'wild goose chase' – think itself competent to judge the extent of reserves in a region that hasn't even been explored? Why does it think it has any right to second-guess the risks energy companies are willing to take?

Even if some of the resources prove to be less rich than antici-pated, the accessibility of unconventional gas reserves still rings the death knell for the environmental movement's alarmism around peak oil. Of course, peak oil theory is in itself questionable. The price mechanism is the means by which the market deals with scarcity. If fossil fuels were to become scarcer or more costly to extract, prices will rise, which will reduce demand and reward investment in better production techniques. If this doesn't help and the scarcity becomes critical, rising prices will make alternatives economically viable. Eventually, the market will switch without any need for government direction or environmental lobbying. After all, we stopped using horses for transport not because we ran out of horses to ride or hay to feed them. We stopped because they were too slow and expensive compared to new technology. Henry Ford once said if he'd listened to public opinion in his day, he'd have concluded that what people wanted was a faster horse.

The threat to the peak oil myth may be the biggest reason why environmentalists are so strongly opposed to natural gas. The availability of an abundant, inexpensive and clean 'bridge fuel' makes a right mess of their pet theory that we're running into a brick wall, and that we'd better get cracking to spend fortunes on expensive, inefficient energy sources such as solar, wind and wave power. Besides, if they can convince government that there's probably not much gas there anyway, they don't have to do the more arduous work of proving that gas exploration and extraction will cause massive, irreversible harm to the environment. As we have seen, their ability to do so is distinctly limited, and relies in large part on anecdotal evidence, wild assumptions and discredited research.

The theory that there's not enough gas to exploit profitably is, of course, patent nonsense. It requires us to assume not only that we can predict a wide range of unknowns, but also that energy companies will willingly drill for gas when it is not profitable to do so. It may not be surprising that environmentalists with scant regard for scientific fact also do not understand elementary economics, but that doesn't make their economic prognostications any more valid.

The reality of the matter is simple: If there is enough gas, energy companies will be interested in extracting it. If not, they won't. We might object if we can show that this is likely to lead to widespread infringement of people's rights or serious environmental damage. To date, the green lobby has failed to do so. Proposing that a government ought to predict whether a project will prove to be profitable before permitting it reveals a distinctly socialist cast of mind. The entire point of private capital owners is to risk capital in competitive ventures. Failures will quickly be eliminated and successes will thrive. Simply put, whether environmentalists believe there is enough gas underground to extract profitably is entirely irrelevant to whether or not energy companies should be permitted to explore for it.

THE EARTH MOVED UNDER MY FEET, THEY FRET

With the exception of occasional anecdotal stories of localised surface spills, which are paraded before the press like so many photographs of air crashes in support of a campaign to ban flying, or pollution incidents that upon investigation turn out to have little to do with hydraulic fracturing, there are a few other arguments that opponents like to raise. One of them is that fracking will cause earthquakes.

As a mining country, South Africa is well acquainted with seismic activity resulting from mining. The entire city of Johannesburg sits atop a warren of mine shafts and tunnels, and tremors are a frequent part of life in South Africa's biggest city. A report on seismicity compiled before the 2010 FIFA World Cup, by AON Benfield Natural Hazard Centre at the University of Pretoria, makes this clear: 'Mining related events add a low magnitude, high frequency facet to earthquake risk in Johannesburg... The tectonic origin and mining related events are considered to be largely uncorrelated.'[14] Human activity does not cause earthquakes, technically speaking. The stresses that produce earthquakes already exist in the tectonic plates where earthquakes originate. These stresses are always likely to be released as a quake at some point in time, but human-induced triggers may cause them to occur sooner than they otherwise would have.

The largest earthquake geologists believe to have been triggered by human activity occurred in 1967 at the Koyna Reservoir in India. Measuring 6.5 on the Richter scale, it caused 200 deaths and significant damage to nearby towns.

South Africa's largest historic earthquakes occurred in 1809 in Milnerton near Cape Town, in St Lucia in KwaZulu-Natal in 1932 and in the Ceres/Tulbagh area in the Western Cape in 1969. All were recorded at magnitude 6.3 on the Richter scale. A magnitude 6.2 quake in Koffiefontein in the western Free State in 1912 was felt across the country.

Earthquakes of more than magnitude 5 occur in South Africa and surrounds about once every couple of years on average. Many are believed to be triggered by mining activities. Klerksdorp measured a magnitude 5.3 quake as recently as 2005. In 1976, a six-storey building collapsed in Welkom, and several subsequent quakes in the mining regions of the Free State and Witwatersrand have caused structural damage. Tremors of magnitude 4 or more are very common on the Reef.

Some of the more significant tremors, like a recent 4.1 magnitude quake centred on the town of Leeu-Gamka on 14 May 2011, originated deep in the Karoo. The Ceres and Koffiefontein earthquake clusters are the most significant to the Karoo region, but other historic quakes affecting the Karoo region were a series of four quakes exceeding magnitude 5 in Sutherland in 1952, two Namaqualand quakes of magnitude 5.5 and 5.8 in 1950 and 1953 respectively, and a far more recent quake of 5.1 on the Richter scale in Brandvlei, in what is now the protected astronomy reserve, in 1994.

Chris Hartnady, a former University of Cape Town geologist with specialist knowledge of southern Africa's structural geology and tectonics, told me that the South African seismic observation network is focused on shallow quakes originating less than five kilometres deep, and scientists cannot easily tell whether quake origins are purely tectonic or induced by mining. He was interested in my own description of the Leeu-Gamka quake as a deep, low-frequency series of thumps, quite unlike the high-frequency rattling that is familiar to Witwatersrand residents, because it confirmed his suspicion that it originated as much as 25 kilometres underground. In a 2009 paper in the *Seismological Research Letters*, Mayshree Singh and others made a similar observation about the limitations of the South African National Seismograph Network.

Despite the frequency of mining-related tremors, South Africa is not exactly an earthquake hot spot. It is in what geologists call a 'stable continental region'. Still, although the risk of earth tremors as a consequence of mining and drilling activity may be minor, they do cause widespread public concern. According to UK-based shale-gas exploration outfit Cuadrilla Resources, several minor tremors, the worst measuring only 2.3 on the Richter scale, may have been caused by a rare combination of geology and hydraulic fracturing. Gas-drilling operations in the region were suspended in the wake of these events, pending investigation. More recently, an increase in quakes, the worst of which measured 4.0, struck Ohio, for much the same suspected reasons.

Once it is known that an area does pose seismic risk, measures can be taken to mitigate it. As it is, authorities in the UK do not expect any tremors of more than magnitude 3 as a consequence of gas drilling, and regulators are instituting measures to prevent even those. Perhaps the best gauge of the real risk is a comment made to me by Adrian Tiplady, of the SKA project, to whom seismic activity would be a real issue:

> Seismic disturbances underneath the telescopes could have a detrimental impact. However, after discussions with Shell, I am led to believe that the size of the cracks is the length of tooth-picks, a hair-width in size, a couple kilometres underground. Based on this, I would consider potential seismic disturbances quite a low risk.

IT WON'T DO MUCH GOOD ANYWAY, THEY ARGUE

Finally, when all the arguments against fracking have crumbled under the weight of factual evidence, critics will lodge their final appeal: promises of job creation are exaggerated and it won't benefit South Africa much to permit gas drilling.

In some ways, this is a moot point, much like it is no business of anyone other than shareholders to evaluate the risk and potential profitability of gas-drilling operations. It is true that many of the actual drilling contractors will be specialists imported from other countries, but then, oil companies have never claimed otherwise. Still, the opportunities are legion for local suppliers to provide everything from technical supplies and hospitality to environmental consulting and legal services. Boom towns in gas-rich regions of the US are testament not only to the minor risks associated with hydraulic fracturing, but also to the economic windfall that gas-drilling projects bring to the community.

A study by the late Tony Twine, an economist with Econometrix, predicted a wide range of potential benefits to the South African economy. Considering both direct and indirect economic activity, successful exploitation of the Karoo gas fields could, he estimated, yield between 300 000 and 700 000 jobs, between R80 billion and R200 billion in GDP and between R35 billion and R90 billion in government tax revenue. 'This is a lot of money, a lot of jobs. To walk away from values like these is going to be a difficult decision,' he told the *Business Day* newspaper.[15]

Equally importantly, in a country struggling with a grave energy shortage, it would be daft to rely entirely on renewable energy dreams. The largest solar thermal power generation units world-wide are rated at a measly 50 megawatts (MW) each. An installation consisting of nine units in the Mojave is the largest operational solar power plant in the world. It generates just over 350 MW. In Spain, a world leader in solar power, several solar thermal power plants of 100 MW and 150 MW are being built. By comparison, the largest photovoltaic solar plant produces less than 100 MW.

Wind farms can in theory produce a little more power, with the largest capable of generating some 700 MW under ideal con-ditions. Of course, rated or 'nameplate' capacity assumes weather conditions that cannot be sustained for anywhere near twenty-four hours a day, 365 days a year, so the actual solar or wind power delivered is significantly lower than manufacturer nameplates suggest.

Nonetheless, such plants may well have a role to play in South Africa's energy mix. South Africa's state-owned electricity monopoly, Eskom, has a stuttering independent power-producer programme, which may soon give investors in green technology the opportunity to put their capital where their mouths are. Com-petition in technology development can only be a good thing.

However, even with the best will in the world, electricity generation on the scale of typical renewable energy plants can deliver only a tiny fraction of the power South Africa needs. Eskom aims to increase capacity by 22 000 MW in the next five years alone. That would require 440 thermal solar-power units, plastered all over the Karoo landscape.

This is true worldwide. Figure 11 shows the history and expected future of the world's energy sources. Renewables do contribute increasing amounts, but for many years we'll have to continue to rely on fossil fuels like oil, gas and coal, whether we like it or not.

According to Philip Lloyd, a research professor at the Energy Institute of the Cape Peninsula University of Technology, a major gas find could transform the economy of South Africa.

Besides creating thousands of jobs, a reliable source of domestic natural gas would permit gas-turbine power stations all along the coast to provide both peak- and base-load power, and enable the growth and development of a great number of industries, from large-scale steel and aluminium smelters to thousands of small factories countrywide.

In addition, there's an opportunity to convert public transport, such as the taxi industry, to run on cleaner-burning natural gas, as well as to generate massive export revenues by exploiting Sasol's world-class gas-to-liquids technology, which at present can produce over 700 000 barrels of liquefied natural gas (LNG) per day, at refineries around the country.

As the gas shortage of 2011 amply demonstrated, a broad swathe of South Africa's manufacturing, hospitality and automotive industries relies on this form of energy. Since gas-to-liquids technology is economically viable with an oil price anywhere north of $20 per barrel, its importance in the present economic climate, with oil nearer $100 per barrel, has been growing sharply. South Africa can be at the forefront of this global trend. Grasping this opportunity would represent a strong shot in the arm for economic growth, which a country struggling with widespread poverty and stubbornly high unemployment can ill afford to pass up.

EXTREME SCEPTICISM

The aim of Chapters 1 to 3 has been to demonstrate, in considerable detail, that almost every claim made by the environmental lobby opposed to hydraulic fracturing is either grossly exaggerated or an outright lie.

During the public debate at Rhodes University, Jonathan Deal, TKAG's national coordinator, admitted that the group's rhetoric was emotional and exaggerated. He claimed that this was necessary to place the debate on the public agenda. The converse of that argument is, of course, that it ought not to be necessary to debate matters that are not true. Public consultation in the policymaking process is, broadly speaking, a good thing. However, a government would be derelict in its duty if it bowed to special-interest lobby groups when they fail to debate matters in good faith.

These errors and persistent exaggerations – uncorrected even when brought to environmentalists' attention – are the grounds on which the green lobby wants the public to base their opinions. This lobby, and in particular the well-funded TKAG, is hell-bent on influencing government policy 'through every legal means at our disposal, including litigation'.[16] It has already launched legal action against the government, although it has done so on a matter of procedure rather than substance. At best, such frivolous legal harassment will delay an informed decision by government, which has wisely called for a moratorium on hydraulic fracturing to give it time to consider fairly the views of all stakeholders.

These stakeholders include not only energy companies and extreme environmentalists. They include Karoo landowners who have been dispossessed of their mineral rights and stand to gain nothing by agreeing to gas drilling on or near their properties. They include business owners who stand to benefit from supplying gas-drilling operations. They include thousands of rural poor, most of them landless and many of them unemployed. The latter have been conspicuous by their invisibility in the public debate, yet they may well stand to gain the most from employment opportunities and economic development in the greater Karoo.

An umbrella body representing a range of union, church, small farmers and community interest groups has been formed under the name Karoo Shale Gas Community Forum, but this group, unlike the environmentalists, has received extremely little attention from the media. It will not surprise readers that this group has expressed its support for fracking. *The New Age* published one of only a handful of news articles that quoted its coordinator, Vuyisa Jantjies: 'We're in support of fracking but the Karoo communities must benefit from the project.'[17]

By overstating its case, and persisting with falsehoods even in the face of public correction, the environmental movement undermines its own credibility. It may well be that certain precautionary measures and safety regulations are necessary. It may be reasonable to extract enforceable commitments from energy companies about taking adequate precautions against accidents, restoring the environment if a spill occurs or compensating farmers for financial losses.

However, when a group sets out with the sole purpose of forever banning any and all gas drilling in South Africa, and supports its call with discredited research and frequent exaggeration, how is it possible to pick out the handful of valid warnings? Why should anyone believe it, when it keeps crying wolf in the most hysterical terms?

If the outcome of this 'debate' is that government places undue restrictions on energy companies, or worse, that energy companies decamp to gas-rich countries like China and Poland, which welcome foreign investment and economic development, the extreme environmentalists will have a great deal to answer for.

When environmentalists openly admit to using emotional rhetoric and exaggeration, and it can be shown that much of their argument consists of a web of carefully constructed, unscientific propaganda, should anyone listen? If these tactics perpetuate poverty and unemployment in a developing country that desperately needs productive industry and economic growth, will South Africans, and indeed the world, just stand by and shrug?

4

Umami
The Good Taste

SILENT SPRING

Chemicals cause cancer. Eliminate all synthetic chemicals from our food and our farms, and we'll live longer, healthier and more prosperous lives, and save the planet while we're at it.

Such is the received wisdom of the modern environmentalist movement.

We should eschew even well-tested allopathic medicine – the kind that works by counteracting the effects of a disease. Instead, we should consume minute doses of homeopathic medicine – the kind that in normal concentrations might work *with* the body, to teach its own defences to fight the disease, much like a vaccine does. (Except, of course, vaccines are also evil and cause autism. But we'll get to that.)

In our food, we should allow no preservatives, no synthetic taste enhancers, no unnatural colourants, or any of the large number of substances that industrial-scale food production has adopted to achieve certain benefits, because the cost to us is getting cancer.

In our environment, we should use no chemical fertilisers, no pesticides, no scientific approaches to breed better plant stocks, no drugs to produce more, tastier or healthier meat, because all of it will somehow make us unhealthy. Or worse.

It all started with Rachel Carson. An American zoologist with a literary streak, she was a moderately famous writer on nature when a series of alarming environmental disasters and food crises highlighted the dangers of the deliberate but careless use of chemicals for pest control. The consequences were often severe by comparison with the intended benefit, and the chloride-based pesticides that had proved so effective against mosquitoes during World War II killed many other insects indiscriminately. In response, Carson, then fifty-five years old, wrote her most famous – or infamous – book: *Silent Spring*, which appeared in 1962.

'There was once a town in the heart of America where all life seemed to live in harmony with its surroundings. The town lay in the midst of a checkerboard of prosperous farms...' she began. Lyrically, she described the wildlife, the flowers, the birds, the streams and the seasons. 'So it had been from the days many years ago when the first settlers raised their houses, sank their wells, and built their barns.'

Whether this idealised vision of the past was accurate or not did not matter – it did not last:

Then a strange blight crept over the area and everything began to change. Some evil spell had settled on the community... Everywhere was a shadow of death. Farmers spoke of much illness among their families. In the town the doctors had become more and more puzzled by new kinds of sickness appearing among their patients. There had been several sudden and unexplained deaths, not only among adults but even among children, who would be stricken suddenly while at play and die within a few hours.

There was a strange stillness. The birds, for example – where had they gone? ... It was a spring without voices.[1]

Soon, we discover the cause of all this misfortune: 'a white granular powder'. Dichlorodiphenyltrichloroethane – or rather DDT, as it is commonly known. 'The people had done it themselves,' Carson laments.

Emotionally disturbing, even though she admits that the account is fictional, this book is widely viewed as the founding impulse, the call to arms, of modern environmentalism. Modern editions sport a bold claim on the cover: 'The classic that launched the environmental movement'. Sadly, its reputation has not survived the last fifty years untainted.

In one corner are the environmentalists themselves, upset by the crass pursuit of profit that they believe drives the destructive behaviour Carson describes. They interpret her book as a dire warning against synthetic chemicals in particular, and interference with nature in general. It was Carson who established the notion that human beings have become too powerful, and in their quest to control nature by means of technical and scientific progress, they are not only *able* to destroy it, but are likely to do so.

In the opposing corner are corporate lobbies for the agriculture and chemical industries, supported by anti-environmentalist conservatives and anti-government libertarians, who see Carson's book as unqualified, hysterical and irresponsible.

Unfortunately, Carson died in 1964 of breast cancer. Fortunately, she did so before the worst accusations against her were levelled: that she was ultimately responsible for the banning of DDT, which caused the avoidable deaths of 50 million or more people who could have been saved from malaria by the judicious use of the pesticide. These critics rank her up there with Stalin, Pol Pot, Mao and Hitler.

Both sides are wrong. Both sides are guilty of exaggeration. And neither side does the rest of us any favours by their shrill extremism.

Rachel Carson was not evil. She raised very real problems, which in the preceding years had several times made newspaper headlines across her native United States.

According to Carson's book, a letter to the *Boston Herald* in January 1958 described the damage in a private bird sanctuary in Duxbury, Massachusetts, after mosquito-control chemicals from crop dusters went astray.

Around the same time, private landowners on wealthy Long Island lost a legal battle to prevent the government from spraying their estates in Nassau and Suffolk counties against a destructive pest known as the gypsy moth, which had caused widespread defoliation in New England's forests.

In 1957, the US launched a large-scale government effort to combat fire ants in citrus plantations of the southern states. DDT proved to be an excellent pesticide against the little fire ant, which so tormented farm workers that agricultural productivity suffered.

Although the side effects were not known at the time, the use of DDT was not exactly thoughtless and indiscriminate. In the US, the President's Science Advisory Committee had determined it to be safe for humans. The government had tested it extensively for efficacy and found it to be superior to existing pesticides, which resulted in only temporary control of the pests in question. The research also found that a spray proved to be more effective than application by brush to affected trees.

One of the dangers of government programmes like these, and why they ended up conflicting with the interests of private individuals, was that the government's research did not extend to the negative environmental effects. Moreover, in the chemical industry the use of pesticides had a great number of supporters. Who wouldn't want to lobby the government to buy enough of your product to spray half the United States? As a consequence the government, never being particularly frugal with tax revenue, bought great quantities of pesticides from its supporters in the chemical industry, and sprayed them as far and wide as politics would allow. When cattle began to die as a result of aerial spraying, however, local farmers balked.

The public backlash began as resistance by private property owners against what they saw as excessive and indiscriminate spraying without concern for collateral damage. Carson took this concern and turned it into a morality tale about nature, which set the stage for a fight between environmentalism and corporate interests.

Like the landowners before her, Carson correctly diagnosed the effect of the pesticide on beneficial species and its accumulation in the ecosystem. But contrary to the claims of her critics, her work did not cause a ban on DDT. Rather, Carson warned that DDT's effects on ecosystems were not fully understood, and that mosquitoes were becoming resistant to the chemical, with the result that continued indiscriminate use of the pesticide only made things worse. (Of course, even that is a simplification of the truth. The phrase 'becoming resistant' is misleading. In fact, mosquitoes with the mutations that made them resistant to DDT always existed, but they gained an evolutionary advantage when their peers started dying from the stuff.)

In particular, Carson did not advocate ignoring insect-borne diseases merely because combating them might require chemical pesticides. What she actually said was: 'Practical advice should be "Spray as little as you possibly can" rather than "Spray to the limit of your capacity".'

That is an important statement. It is a reasonable evaluation of risks versus benefits, not an extreme exaggeration. It contrasts strongly with the everyday advice of modern environmentalists – namely, to eliminate potentially harmful chemicals altogether.

Not only are the claims of a 'ban' overstated, but so were the reasons for the discontinuation of DDT use in many cases. The US did impose a ban on agricultural use in 1971, and the Stockholm Convention did the same worldwide in 2001, but in both cases exceptions were made for disease control. Among the reasons why DDT fell into disfavour even before the US ban was that Carson's prediction of mosquito resistance was all too true.

Members of the movement inspired by her work, however, adopted her emotional rhetoric and took it to extremes. They idealised the world untouched by modern industry that Carson described so eloquently, and feared the horrors that befell the town in her narrative, even though she explicitly stated that this town was fictional. Concerned readers – and who wouldn't be concerned over their health, their children's well-being and the state of their environment? – took to heart her warnings that chemicals may have unintended consequences, including environmental damage and severe health risks, such as cancer.

They read about the effect DDT has on the eggs of the peregrine falcon. Affected birds who ate poisoned prey laid infertile eggs, and even their fertile eggs were in danger because of thinning shells. However, this was anecdotal evidence, gathered because this bird happened to be easy to study. Readers were led to believe – wrongly – that this didn't happen just to some birds in some circumstances, but that DDT had this effect on all birds.

Thus, a small and manageable risk became established in the public mind as a *certain* danger that must be avoided. The popular notion of chemical pesticides that accumulate in the tissues of animals and find their way through the entire food chain, killing and maiming as they go, was born. By the 1970s, this narrative was explained in picture books for young children.

Carson offered another anecdote, about a woman who came into contact with DDT, contracted leukaemia and died within three months. Readers were, once again, left free to conclude that DDT causes cancer, with the same certainty that they know smoking causes cancer. That their icon succumbed to cancer herself merely underscored the urgency of her warning.

Among environmentalists, this view has been elevated to a dogma that contrasts the noble pre-modern human being living healthily in an untouched environment with the wholesale destruction and poisoning of the planet by faceless corporate greed. It is a dogma against which Rachel Carson herself might have objected, had she lived.

Fifty years later, it is clear that predictions of a 'silent spring' were overdone. No doubt environmentalists will wish to take credit for this fact, but the truth is that Rachel Carson presented partial evidence, some of which was anecdotal at best, and her supporters in turn exaggerated her work. Ranged against them, and inadvertently lending the environmental narrative much credence, were detractors with vested interests, who exaggerated the terrible harm she supposedly caused. The acrimony between the two sides runs deep, and accusations of dubious motives and unsavoury connections continue to this day.

The truth is that the backlash against DDT did have positive effects, since indiscriminate aerial spraying did do very real harm. However, its disuse also had significant negative consequences.

For example, the gypsy moth remains a pest. It was introduced to the US in the nineteenth century by a French entomologist, Étienne Trouvelot, in an attempt to create a domestic silkworm industry. In their native Europe, gypsy moths had been blighted by disease, so the idea that they could become a pest was far from his mind.

The insect thrived, although attempts to harvest its silk proved not to be commercially viable. How the moths escaped from Trouvelot's custody no one will ever know, but once free, the larva proved to be a highly destructive caterpillar. It is an unfussy eater, and although it prefers certain deciduous trees – including valuable timber species – it will eat almost anything when necessary. It has caused catastrophic defoliation throughout an ever-growing region. According to the US Department of Agriculture, more than 340 000 km^2 – an area more than half as large as the Karoo Basin – have been defoliated by the gypsy moth since 1924.

Initial control measures were labour-intensive and, although partly successful, never did prevent the pest from repopulating infested areas. In 1956 and 1957, almost 15 000 km^2 were sprayed with a mixture of kerosene and the broad-spectrum chemical pesticide DDT. The question that emerges is whether the discontinuation of the use of DDT, for all the benefits that might have had, did not also cause serious harm.

To this day, periodic gypsy moth outbreaks continue. Research has led to the development of a biological pesticide known as Bt (for *Bacillus thuringiensis*, a bacterium that affects the larvae), which, while less destructive than DDT, also affects other butterflies and moths. Currently in place is a comprehensive and very expensive control programme involving a combination of techniques, including nineteenth-century-style manual removal of eggs, pheromone-baited traps and biopesticide application.

Despite millions invested in research and the availability of modern technology, the gypsy moth continues to pose an alarming threat, not only to the original infestation areas of the north-eastern US, but also to similar forests and woodlands further down the east coast and on to Missouri. The most recent campaign against the pest is titled, dispiritingly, 'Slow the Spread'.

Likewise, malaria continues to pose a grave threat. In sub-Saharan Africa, it is the leading cause of death in children under five years of age, according to Africa Fighting Malaria, a non-governmental organisation (NGO) advocating the judicious use of DDT. Of as many as half a billion cases of malaria each year, a million result in death. This is the basis for the claim that Carson killed 50 million people with her book since it came out fifty years ago.

Contrary to frequent newspaper headlines and films such as Al Gore's *An Inconvenient Truth*, malaria is not on the increase because of climate change. Areas highlighted in such work for their malaria risk are no strangers to malaria, as Paul Reiter, a professor of medical entomology at the Pasteur Institute in Paris, notes. A member of the World Health Organization (WHO) Expert Advisory Committee on Vector Biology and Control and a specialist in mosquito-borne diseases, Reiter wrote an editorial in which he complains:

> For 12 years, my colleagues and I have protested against the unsubstantiated claims that climate change is causing the disease to spread. We have failed miserably ... We have done the studies and challenged the alarmists, but they continue to ignore the facts ... Malaria's return in the past 20 years has been due to many factors – the effective ban on DDT, deforestation, migration from highly malarious areas, drug and insecticide resistance and above all, poverty.[2]

DDT remains the most effective insecticide for the control of malaria mosquitoes, and is listed at the top of the WHO's list of recommended chemicals. It is the only insecticide that remains effective for longer than six months.

Today DDT is an integral part of the global campaign against malaria. This struggle may not have been quite so arduous if it hadn't been for exaggerated fears about the dangers of DDT among environmentalist elites in the West, after the last Western country was conveniently declared malaria free in 1971.

THE DANGEROUS MYTH ABOUT VACCINES

With parents having been primed to fear chemicals, it took just a single scientific study to turn many of them against common vaccines. In fact, the fear of so-called toxins in vaccines given to children has existed ever since the first vaccines against smallpox made their appearance.

In the eighteenth century, smallpox was the reported cause of 20 per cent of all deaths in London, according to the National Network for Immunization Information (NNII). This non-profit organisation counts among its affiliated members the Infectious Diseases Society of America, the Pediatric Infectious Diseases Society, the American Academy of Pediatrics, the American Nurses Association, the American Academy of Family Physicians, the National Association of Pediatric Nurse Practitioners, the American College of Obstetricians and Gynecologists, the University of Texas Medical Branch, the Society for Adolescent Medicine and the American Medical Association.

'Mothers counted their children only after they had had the smallpox,' ran an expression at the time, before Edward Jenner took a centuries-old Chinese practice called 'variolation' and developed it into what we would recognise today as immunisation. The Chinese had discovered that by intentionally infecting people with a small amount of smallpox-carrying discharge from victims of the disease, they were likely to contract a mild form of it, before recovering and becoming immune to future infection.

Jenner observed that dairymaids were often immune to smallpox, thanks to their exposure to a milder relative of the disease, cowpox. In 1796 he used material from the lesions of infected cattle to perform the first 'vaccination'. The word comes from *vacca*, which is Latin for 'cow'. It was gross, but it worked.

Vaccination became law in the West during the nineteenth century, and spread worldwide in the twentieth century. The last reported case of smallpox occurred in Ethiopia in 1976, and in 1980 scientists declared the disease eradicated. That smallpox is no longer a threat to modern populations is due entirely to the development and rigorous universal administration of modern vaccines.

Polio has been eradicated in the Western hemisphere, and many other diseases have declined sharply since vaccinations against them began. According to the NNII, 'diphtheria declined from a high of 206,939 cases in 1921 to just one in 1998; whooping cough declined from 265,269 cases in 1934 to 6,279 in 1998; and measles has fallen from 894,134 cases in 1941 to just 89 in 1998'.[3]

The list doesn't end there. Hepatitis A and B occurred over 26000 times each in the US when vaccination against them began in 1995 and 1986 respectively. By 2006, incidence for both was down to fewer than 5000 cases. Ninety-nine per cent of the millions of annual cases of chickenpox have been eradicated in the US by just over ten years of vaccination. Rubella is down from almost 50000 cases in 1969 to just eleven a year in 2005 and 2006. Other diseases successfully fought with vaccines include anthrax, rabies, tetanus, shingles, yellow fever, human papillomavirus (a potential precursor to cervical cancer), Lyme disease, meningococcal and pneumococcal disease, rotavirus, typhoid fever and tuberculosis.

Ironically, the fact that vaccines work means that the diseases against which they immunise children aren't often seen. Therefore, the fear of those diseases is often less than the fear that a child may develop an adverse reaction to the vaccine.

That fear has recently been stoked by one of the most egregious cases of fraudulent science and media alarmism on record. A study by Andrew Wakefield in 1998 tested the theory that there might be a causal connection between the measles, mumps and rubella (MMR) vaccine and childhood autism. Although the published study said that no such connection could be demonstrated, he gave interviews to the media suggesting that there was a strong case for such a link, and that the British authorities had failed to properly test the vaccine for safety.

This warning spread rapidly among those who fear chemicals in their food and medicine, and celebrities took up the cause. *Playboy* Playmate Jenny McCarthy and her then boyfriend, actor Jim Carrey, funded an organisation known as Generation Rescue, which decries the poisonous contents of vaccines. As recently as 2008, the organisation ran full-page adverts in newspapers trumpeting the MMR–autism link.

A typical tactic of such groups is to list the ingredients of a particular vaccine and then describe each chemical's effects. Many of these descriptions include warnings that the substance in question is carcinogenic.

Of course, such a rhetorical tactic is fundamentally misleading. Most medicines are therapeutic in small doses and poisonous in large doses. Citing the effects of a large dose as the likely consequence of small amounts is, quite simply, exaggeration.

Some of the most extreme examples found by David Gorski of the blog *Science-Based Medicine* occur in the work of Shauna Wood, writing for *Vaccination News*. 'Hydrochloric acid,' she writes, 'CAN DESTROY TISSUE UPON DIRECT CONTACT! Found in aluminum cleaners and rust removers.' She also notes: 'Sodium Hydroxide (also known as lye, caustic soda, soda lye.) Is corrosive and is an Eye, skin and respiratory irritant. Can burn eyes, skin and internal organs. Can cause lung and tissue damage, blindness and can be fatal if swallowed. Found in oven cleaners, tub and tile cleaners, toilet bowl cleaners and drain openers.'[4]

If this sounds like the alarmism about fracking chemicals, that's because the tactic is just the same. This truly is scary stuff. Worse, it is all true. I'd strongly advise against drinking bottles of either substance, or injecting them into your babies. However, if you mix the two, as one does in order to neutralise solutions that are too acidic or too alkaline, one gets a solution of common table salt. That, even in large doses, is not going to rip your lungs out, remove rust or clean your toilet. In fact, saline solution is a very cheap and harmless substance that is commonly used in intravenous drips for rehydration therapy or for cleaning sensitive areas of the body, like the eyes and mucous membranes, affected by irritants.

With the MMR–autism study, however, the alarmists had an apparently more credible and more serious piece of 'evidence'. It was published in the highly respected medical journal *Lancet*, survived superficial scrutiny by people who passed high school chemistry and was described in exaggerated terms by its author.

Much like with the Cornell and Duke fracking studies discussed in Chapters 2 and 3, Wakefield's study stood as the sole basis for the claim that vaccines could have severe health consequences in children – until Brian Deer came along in 2004. A journalist for the British *Sunday Times*, he launched an investigation, documented in detail on his website, which uncovered the truth: Wakefield and his dozen associates were wrong. Not only were they wrong, but Wakefield had been paid £400 000 by the plaintiffs in a prospective class-action lawsuit against vaccine manufacturers to find as he did, and he doctored the results to achieve what they wanted to prove.

As a result of Deer's sleuthing, the *Lancet* first tried to bury the story, then issued a partial retraction in 2004 and finally retracted the paper in its entirety in 2010. But for twelve years, the false story had been used to needlessly scare the public.

The psychology of such a scare was instructive, too. Bill Ahearn, director of research at the New England Center for Children, a private non-profit educational facility for children with autism, reported that '[a study in 2002] found that parents of children with autism diagnosed after the MMR controversy was publicized in the media were more likely to report the onset of autism as just after MMR vaccination than were parents of children with autism diagnosed before the controversy'.[5] In other words, because they were told the vaccine could cause autism, they blamed the onset of autism on the vaccine. Ahearn directly blames sensational reporting for the sharp decline in MMR vaccinations since the scandal broke more than a decade ago:

> The impact of the media's coverage of this issue has had a significant and detrimental influence. Unfortunately, highly improbable events, extraordinary claims implying a conspiracy, and steadfast beliefs with little support beyond anecdote tend to be given more coverage than sound information based upon empirically valid and peer reviewed research.

Most anti-vaccination campaigners no longer cite the study, but some, including Wakefield himself,[6] try to rubbish Deer's work as a smear campaign by the Goliath 'Big Pharma' to discredit a brave David.

It isn't difficult to find parents who, having once believed the MMR–autism claim, are now embarrassed to have been gulled and continue to raise objections. The objections usually consist of anecdotal tales of side effects that appear after a child has been vaccinated. Aside from the obvious statistical problem of citing anecdotes as if they indicate the degree of risk, such arguments suffer from the *post hoc ergo propter hoc* fallacy: just because one event happens after another does not mean that the first event is the cause of the second. Other cases in which the effect occurs may not have been documented, the effect may have been caused by something else, or a third cause may underlie both observed effects. In any case, vaccinations declined sharply in the last decade and preventable diseases mounted a resurgence, according to Ahearn. Most of these occur in the developing world.

India accounts for 69 per cent of the world's measles deaths in children under five, a fact which a case study in the science journal *Nature* attributes to a lack of money and political will to roll out vaccination programmes. Southern Africa makes up most of the rest of measles deaths in childhood, with 21 per cent. The rest of the world combined accounts for only 10 per cent of all childhood measles deaths.

In Africa, twenty-eight countries experienced measles outbreaks in 2010/11, in part as a consequence of public opposition to vaccination. In Europe, 30 000 measles cases were reported in 2010, five times more than the annual average in the five years before. Among the causes cited by *Nature* are the unfounded fears about the MMR vaccine. In the United States, measles had officially been eliminated in 2000, but new cases were imported, and by 2011 more cases were registered than in 1996, leading *Nature* to cite 'fears of outbreaks among unvaccinated children'.[7]

Wakefield's medical licence has been withdrawn and his paper has been retracted, but the damage has been done. And as with most examples of exaggeration about pollutants or chemicals, the heaviest burden in terms of human lives and living standards falls on developing countries, which lack the abundant resources to combat public disinformation caused by exaggerated fears or to mount new, broad-based public health initiatives.

THAT FINGER-LICKING TASTE

Chemicals in food are unfashionable. The reason, for the most part, is the fear that they contribute in some way to allergies, as well as more serious diseases such as cancer. Eating only 'organic' foods, as nature intended them, is the safest way to eat, so the common wisdom goes.

Among food additives, few chemicals have a worse reputation than monosodium glutamate, or MSG. It allegedly causes 'Chinese restaurant syndrome', a severe allergic reaction.

Ironically, however, MSG's reputation among food additives is perhaps the least deserved. The term might sound awful, but then so does sodium chloride (table salt) or omega-three fatty acids (which are far better for you than omega-six fatty acids).

All of these sound like the kinds of chemicals that cause our rising cancer rates. Yet the most likely way they will kill you is by worrying yourself to death. Just because a name sounds 'chemical' doesn't mean it's dangerous.

The fabulous magic and comedy duo of Penn & Teller, who often use their acts to satirise or debunk myths, once drew up a petition to ban a chemical substance. They sent out a pretty girl to collect signatures and explain to people that this substance was in all our reservoirs, lakes and streams; was used in pesticides and fertiliser; and ended up in our grocery stores, meat and baby food. It caused excessive urination and perspiration, among other evils. The chemical in question was dihydrogen monoxide, and it ought to be banned.

The website *DHMO.org* points out that this substance plays a role in the formation of Hodgkin's lymphoma, Ewing sarcoma, chondrosarcoma, fibrosarcoma, multiple myeloma, colorectal cancer, leukaemia, basal cell carcinoma, squamous cell carcinoma and malignant melanoma. If so, it surely ought to be banned.

Dihydrogen monoxide has the chemical formula H_2O. It is water.

The alacrity with which fearful people, conditioned by years of environmental exaggeration, signed the Penn & Teller petition was preserved on YouTube, where, at the time of writing, 2.5 million viewers have had a chuckle at their stupidity.

So what is this monosodium glutamate, and should it scare us?

In the early twentieth century, Kikunae Ikeda, a Japanese scientist from the University of Tokyo, travelled to Germany for post-graduate study. He was particularly interested in organic chemistry, and was intrigued by the tastes of various foods he found in Europe. At the time – and I was taught this myself in primary school – it was believed that there were four basic tastes, associated with four regions of the tongue. These were sweet, salt, sour and bitter.

Ikeda begged to differ. As he later told an audience at Chicago's Eighth International Congress of Applied Chemistry of 1912: 'An attentive taster will find out something common in the compli-cated taste of asparagus, tomatoes, cheese and meat, which is quite peculiar and cannot be classed under any of the well defined four taste qualities, sweet, sour, salty and bitter.'[8] What he found might be described as savoury, or tangy, or full-flavoured. He described this good taste as *umami*, the Japanese word for 'tasty'.

The taste of food was, at the time, a subject of worldwide research, as scientists sought inexpensive but palatable foods to sustain growing but poor populations. In Germany, it is likely that Ikeda encountered Justus von Liebig, one of the founders of the field of scientific food research, and who developed the concentrated extract that became known as Oxo beef stock cubes. In any case, Ikeda had certainly read an article by another Japanese scientist, Miyake Hide, which claimed that better-tasting food helped diges-tion. This inspired in Ikeda the idea of manufacturing an inexpensive seasoning that would improve Japan's poor diet by making bland but nutritious food more enjoyable.

Back home, Ikeda realised that this same taste was noticeable in the traditional Japanese seaweed broth known as *kombu dashi*. For centuries, the Japanese have used seaweed to improve the flavour of food. Ikeda was the first to identify and isolate the ingredient in seaweed that produced this effect: glutamate. Its most common form is a salt, monosodium glutamate.

In the early twentieth century, Japan was modernising rapidly and became fertile ground for a junction between scientific discovery and mass manufacturing, as this dovetailed neatly with the then fashionable concern for the health of the nation. Ikeda approached the Suzuki Chemical Company to produce his newly patented substance. A division known as Ajinomoto (literally, 'essence of taste') began to produce MSG commercially in 1909.

At first, MSG was produced by laborious extractive techniques. It truly was a simple plant extract, and as natural as you please. In 1956, a fermentation process was invented that could produce glutamic acid. More recently, techniques were developed to produce the same or similar flavour enhancers by hydrolysing proteins to produce glutamate or its derivatives.

Glutamic acid is one of the twenty amino acids making up proteins. It is a non-essential amino acid, in that the body is capable of producing it in its free form. (Essential amino acids must be taken in as food, since the body cannot produce them.) While bound glutamate forms part of proteins, free glutamate found in food (such as MSG) acts as the primary energy source for the intestine while digesting food. Ikeda's contemporary was quite correct in surmising that good taste aided digestion.

Glutamate or its salts occur naturally in some foods, such as meat, tomatoes, some seafood, Asian seasonings such as soy sauce and fish sauce, and Italian foods such as anchovies and Parmesan cheese. It does for savoury foods what fat does for meat: it makes it taste better. When you grill a steak, one that contains a little more fat will be tastier, richer and more full-flavoured. A lean steak may be healthier, but with the fat you sacrifice taste. This is why it is much nicer to fry or grill food with a little butter, fat or oil.

Glutamate even occurs in breast milk, and a newborn infant consumes up to ten grams of bound glutamate and one gram of free glutamate per day. Cow's milk contains ten times less glutamate. Only 4 per cent of ingested glutamate enters the body through the intestine, and toxicology tests on whether normal ingested doses cause adverse reactions have yielded inconclusive results. High doses of intravenously administered MSG does cause transient symptoms such as headache, flushing and nausea.

Critics of MSG, such as the Truth In Labeling Campaign, claim that no such safety tests have ever been done, but WHO data on MSG toxicology in humans – complete with real human subjects – go back to at least 1970, which suggests that claims about its untested nature have been untrue for at least forty years.

Katherine Zeratsky, a nutritionist at the prestigious Mayo Clinic, sums it up as follows: '[R]esearchers have found no definitive evidence of a link between MSG and [a range of anecdotal] symptoms. Researchers acknowledge, though, that a small percentage of people may have short-term reactions to MSG. Symptoms are usually mild and don't require treatment.'[9]

The reactions are often associated with Chinese food, because a Chinese immigrant named Robert Kwok, who at the time was a senior research investigator at the National Biomedical Research Foundation in Silver Spring, Maryland, first identified it in 1968. Doctors know it as Kwok's disease, and Kwok himself gave it the title 'Chinese restaurant syndrome'.

That name, too, is a fallacy. Aside from the fact that MSG is a traditional Japanese flavour enhancer, the European Food Information Council reports:

Despite a small number of persons reporting sensitivity to mono-sodium glutamate, scientific studies have not shown any direct link between monosodium glutamate and adverse reactions. Monosodium glutamate used to be blamed for the 'Chinese Restaurant Syndrome' because the first anecdotal report was made following consumption of a Chinese meal and monosodium glutamate is widely used in Asian cooking. Symptoms said to be experienced included burning sensations along the back of the neck, chest tightness, nausea and sweating. However, a double-blind controlled challenge of individuals claiming to suffer from the 'syndrome' failed to confirm monosodium glu-tamate as the causative agent. Other studies have found that allergic-type reactions after Asian meals are more often due to other ingredients such as shrimp, peanuts, spices and herbs.[10]

Unable to prove that MSG is dangerous, groups like Truth In Labeling resort to claims about 'slow neurotoxins'. They have little more than speculation to back up such claims, but it's a scary way to say 'we can't demonstrate that it will kill you now, but it will kill you one day, mark my words'. Those well versed in the ancient art of prophecy will recognise the technique of 'laying a curse'.

The irony of the exaggeration about the dangers of chemicals and the virtues of 'natural foods' is that many additives have dis-tinct beneficial effects. MSG permits people to reduce their salt intake, and that is proven to be good for your health.

Moreover, Ikeda's original purpose for MSG – increasing the palatability of inexpensive but nutritious food and improving the digestion of it – remains an important goal in developing countries, where food isn't as abundant as it is in Europe or the United States.

In the rich world, MSG can reduce the intake not only of salt, but also of fat, which has proven health benefits. Aspartame, another product of the Ajinomoto company, can substitute for sugar, which is a leading cause of obesity.

Preservatives might not be necessary in a world in which everyone has a fridge, nobody spends more than a small percentage of their income on food and it is routine to stop at a fully stocked supermarket on the way home from work. In the poor world, however, many consumers, even in relatively rich developing countries such as South Africa, live on incomes of $50 a month or even less. They spend a third of it on food, and many do not have access to refrigeration or even the electricity to power it. Gastro-enteritis caused by spoiled food is a scourge that kills many children in the developing world.

Traditionally, preservatives such as salt, sugar, vinegar and alcohol were used to deal with these problems. Less expensive, more effective and healthier modern preservatives can significantly reduce food waste, lower the cost of food and cut the risk of getting ill from consuming stale food.

Flavour enhancers such as MSG are even more harshly condemned because of the supposedly lesser importance of taste over shelf life. The benefits of a pleasant life and enjoying one's food might strike one as particularly important, but not so for those who denounce MSG as a harmful food additive used by greedy multinational corporations to exploit our weaknesses for their profit.

If a multinational corporation is greedy enough to use ancient Japanese lore in the effort to sell us tasty food that won't make us sick, we ought to be quite happy to part with our hard-earned money. It is fashionable but misguided to exaggerate the dangers of doing so, and it actively harms the world's poor.

Of course, the paranoiacs who promote expensive, bland health food and frown on anything that sounds remotely chemical in nature – not realising that everything under the sun is chemical in nature – would respond to all this by saying that the studies are flawed, the food safety regulators are in league with big business, and besides, aren't we all dying of cancer?

Let's analyse the cancer claim itself then, since that is the underlying fear, whether it involves pesticides, drugs or food additives.

CANCER WILL GET YOU IF NOTHING ELSE DOES

Cancer is the most common reason for our distrust of 'chemicals' in food, agriculture or the environment in general. But are people really dying more often of cancer than they used to? If so, our efficient agriculture, effective drugs and tasty, plentiful food may account for it. If not, then why is the fear so commonly raised in our media and at our dinner tables?

According to Egyptologists Rosalie David and Michael Zimmerman, writing in the journal *Nature*, cancer is second only to cardiovascular disease as a cause of death 'in industrial societies'.[11] In response to the discovery of traces of cancer in a mummy from ancient Egypt, they wrote: 'A striking rarity of malignancies in ancient physical remains might indicate that cancer was rare in antiquity, and so poses questions about the role of carcinogenic environmental factors in modern societies.' They blame 'modern lifestyle issues', such as tobacco and industrial pollution, for the apparent rarity of cancer in ancient times compared to how common it appears to be today.

Paul Ehrlich, the man *Life* magazine once dubbed 'ecology's angry lobbyist', is infamous for his 1968 prediction that mass starvation in the 1970s or 1980s would claim four billion lives. Ehrlich also penned an article in 1970 titled 'Eco-Catastrophe!' In it, he foretold that the ecological consequences of persistent chemical insecticides and fertilisers would reduce Americans' life expectancy to a mere forty-two years by 1980.

Basing his environmental activism on little more than his own study of butterfly populations and Malthusian mathematical models, Ehrlich was a master at fanning the flames of our deepest fears. Like much of the nascent environmental movement of the day, he was inspired by the dire warning of Rachel Carson's lament in *Silent Spring*. However, his prophecy of doom, like most of his other predictions, proved to be wrong. Not just a little bit wrong, but spectacularly so.

By 1980, life expectancy in the US had not plummeted, but had increased from 70.8 to 73.8. When Ehrlich made his prediction, 158 countries reported life expectancies higher than forty-two. By 1980, 171 countries exceeded Ehrlich's prediction for the US. Today, US life expectancy is an astonishing 78.2, and only Mozambique, Angola and Haiti rank below Ehrlich's dark prediction.

In the modern era, cancer rates did rise for some time until around 1990. However, in 1900, the three most common causes of death were influenza, tuberculosis and intestinal diseases. Today, two of these rarely cause death, while influenza has become far less deadly. The life expectancy at birth in England was well below forty for all of history since records began around the year 1200, and only began to rise noticeably in the middle of the nineteenth century. By 1900 it was still below fifty, but by 1950 it crashed through seventy, on its way to today's eighty.

Notes Danish statistician Bjørn Lomborg in *Newsweek* magazine: 'Globally, life expectancy today is 69. Compare this with an average life span of 52 in 1960, or of about 30 in 1900. Advances in public health and technological innovation have dramatically lengthened our lives.'[12]

So instead of dying of gastro-enteritis as a child, or of the common flu as a thirty-year-old, and becoming a wise old village elder at forty, we carry on living. Don't tell Ehrlich this, but these days life begins at forty. To rub salt into the wound, the only major disease that has noticeably affected life expectancy since 1970 is AIDS, which is clearly not caused by human economic development.

Much of the science and popular literature of the age repeated this alarmism, however. Dystopian fiction writers and magazine journalists vied to give colour to the dire predictions of a future blighted by environmental destruction and deadly disease.

Despite being spectacularly wrong, the notion that we're getting more cancer because of chemicals in our agriculture, medicine, food preparation and general environment is persistent. Mention the word 'cancer' in general conversation, and someone will solemnly opine that instead of increasing quality of life and health, modern life actually causes cancer. They'll warn darkly about the dangers of antiperspirant spray, or mobile phones, or hair dye, or food additives, or – I kid you not – milk.

It is true that some substances are known carcinogens, but none of these is among them. Carcinogens in the environment are a risk, but only one in fifty cancer deaths in the developed world can be attributed to environmental pollution caused by industrialisation.

The problem is that we're living longer. This exposes us to age-related conditions such as heart disease, stroke and cancer. According to the American Cancer Society, your chance of contracting cancer during your first forty years is about one in sixty, but the chance of getting it after forty is one in three.

Cancer incidence did indeed rise for most of the last century, but if you adjust for ageing, cancer incidence was no higher by the end of the twentieth century than at its midpoint. If you also adjust for smoking, as Bjørn Lomborg did in *The Skeptical Environmentalist*, cancer mortality is down almost 30 per cent in the developed world. Despite the fact that new cancer risks do exist, the catastrophic epidemic Ehrlich and Carson foretold is pure fear mongering.

Another reason cancer appeared to become more prevalent during the twentieth century is not because modern life is more taxing on our bodies than the 'natural' lives our ancestors lived, but because modern medicine is better at diagnosing it. Back in the day, many cancers weren't diagnosed, even if they turned out to be lethal. As recently as 1926, a Nobel Prize was awarded for the 'discovery' that roundworms cause stomach cancer. They don't. Today, doctors know a lot more, and use far better equipment. Widespread screening programmes are designed to find cancers – especially breast and prostate cancer – early, cases which might otherwise have gone unreported. No wonder we see more cancer.

Treatment has also improved dramatically. Many people now survive cancers that would have killed them a few decades ago, or live many more years of quality life than their parents could reasonably have expected with the same condition. In the US, the five-year survival rate of all cancers has increased from roughly a third in 1950 to almost two-thirds fifty years later.

Even the data are no longer comparable to those of Ehrlich's day. They show higher incidence of cancer, but they're based on a 'standard population' for the year 2000, which is considerably older than the 1970 standard population used by US health authorities for age adjustment.

As Robert Weinberg, a cancer researcher at the Whitehead Institute for Biomedical Research in Massachusetts, put it to George Johnson for a *New York Times* article:

> There is no reason to think that cancer is a new disease. In former times, it was less common because people were struck down in midlife by other things ... We now diagnose many cancers – breast and prostate – that in former times would have

remained undetected and been carried to the grave when the person died of other, unrelated causes.[13]

And what of the period after 1990? Well, the news only gets better. According to the American Cancer Society, both the incidence of and the death rate from cancer have been declining. This trend is evident in four of the five major cancer types: lung, prostate, breast and colorectal cancer. (Only liver cancer is an exception, which occurs largely as a result of smoking and drinking.)

According to the WHO, the leading causes of these cancers are obesity, low levels of fresh fruit and vegetables in the diet, lack of exercise, smoking and drinking. Note the absence of MSG and DDT from this list. Nor do they lurk just outside the leading causes. The rest of the list offered by the WHO consists of sexually transmitted infection with the human papillomavirus, urban air pollution, and indoor smoke from the household use of solid fuels.

If modern medicines, food additives or agricultural chemicals are so bad for us, it sure doesn't show up at the hospitals.

As we have seen, the risks of chemicals in modern life appear to be overblown, as is the scale of the supposed cancer epidemic itself. In truth, economic development and scientific advances have combined to make us all healthier. When we or our friends or family members die of cancer, it is rarely because of chemicals in our environment. Usually it is because they didn't die at a younger age of gastro-enteritis, smallpox, polio, influenza or malaria. Because so many infectious or contagious diseases no longer kill us, we now succumb to degenerative diseases. And even those are on the decline in prosperous countries.

Despite these facts, the dangers of chemicals and the risk of cancer are routinely exaggerated by the environmental movement. We have touched on some of the costs to society – and especially to developing countries – and we'll see more evidence of such exaggeration, and the harm it does, in due course.

5

Many People Died, But Not at Fukushima

A DAY OF HORROR

Friday 11 March 2011 seemed a day like any other. In Johannesburg and Cape Town, the quarter-to-eight morning traffic was bumper-to-bumper.

In Austin, Texas, it was a quarter to midnight the previous evening. I had just arrived at a downtown hotel, exhausted after several delays on my flight from South Africa, to attend South by SouthWest, the massive film, music and technology conference that Austin hosts every year.

In Japan, it was 14:45 on an ordinary Friday afternoon.

A minute later, disaster struck. About 130 kilometres off Tōhoku, on the eastern coast of the Japanese island of Honshu, an earthquake of magnitude 9.0 struck. The deadly damage was felt as far away as Tokyo, 380 kilometres to the south-west, but it got worse. The quake also spawned a monster tsunami, the likes of which has seldom been seen.

By the time the bodies were counted, 19 214 people were confirmed dead or missing.

There is a reason the English-speaking world uses the Japanese word 'tsunami' for the massive shock wave on the ocean surface that races away from the epicentre of an earthquake at speeds of up to 800 km/h. Japan's island archipelago, situated on the Pacific 'Ring of Fire', where the earth's most violent tectonic-plate activity occurs, is no stranger to the phenomenon. Japan experiences a tsunami about once a year, of which one in five generates significant waves of over two metres.

The earthquake that hit that Friday afternoon was not every year's earthquake. Only four earthquakes have ever been measured that were as strong or stronger, and this was the strongest ever to hit Japan. All the others occurred around the Pacific Rim, too. The record is held by Puerto Montt in southern Chile, which was struck by a magnitude 9.5 quake in 1960. Alaska recorded a 9.2 quake in 1964. The great and terrifying Indian Ocean tsunami of 26 December 2004 was caused by an earthquake measuring 9.1 on the moment-magnitude scale (which has replaced the old, better-known Richter scale), and another 9.0 quake hit the Kamchatka Peninsula of Russia in 1952.

There is no direct connection between earthquake severity and the size of the ensuing tsunami, but at a maximum water height of 38.9 metres (according to the National Oceanic and Atmospheric Administration database), the Tōhoku event also reached the top-ten list of the highest tsunamis in recorded history.

This list includes exceptionally localised events such as the Lituya Bay rockslide in Alaska, which caused a tsunami of 525 metres, but killed only five locals and had dissipated to 30 metres by the time the wave reached the outlet of the bay. Japan itself has seen higher tsunamis only twice before, in 1771 and 1993.

In terms of death toll, only five tsunamis have been worse, and only one of those occurred in living memory. The Indian Ocean tsunami in 2004, because of its wide impact area, took an estimated 227 000 lives. Japan has not seen a tsunami of quite this tragic magnitude since the Sanriku tsunami of 1896.

In the 2011 tsunami, more than 120 000 houses were destroyed and another 200 000 damaged, and total costs of the damage were estimated to be between $200 billion and $300 billion, which makes it the most expensive tsunami in history by an order of magnitude.

Some 2.5 million people were left without water supply, and almost a million were stranded without electricity. Even ten days after the tsunami, over 600 000 people along the north-eastern seaboard of Honshu remained without electricity, the situation having been exacerbated by incompatible electricity networks in the northern and southern halves of the island. This failure of basic infrastructure severely hampered rescue efforts. Reconstruction, according to experts, could take five years or more. This was no ordinary event, and the consequences were severe.

Could anything overshadow the deaths of almost 20 000 people? Or the staggering cost to Japan's economy?

In the event, something did. It was an event that claimed not a single life, but it dominated the press releases of environmental organisations and the headlines of newspapers for the rest of the year.

There's a reason the tsunami event is not generally remembered by the name Tōhoku, or Honshu, or even Great East Japan Earthquake. Less than a year later, the disaster is remembered among most of us by one name only: Fukushima.

ENTER THE NUCLEAR DRAGON

Within hours of the tsunami, as the damage reports piled up, the news took a turn for the sensational. It took environmentalists several weeks to pin the severity of the 2004 Indian Ocean tsunami on humans (the loss of coastal mangrove swamps did us in). In Japan, it was different. Here, environmentalists had a gift horse, and they weren't going to look it in the mouth.

On a stretch of the Japanese coast that took the heaviest battering from the waves stood a complex of venerable old nuclear reactors. They were part of the nuclear energy network that Japan had been building since the oil shocks of the 1970s, which reduced its reliance on imported oil from 77 per cent in 1973 to 43 per cent in 2009, and at the time of the tsunami supplied about 30 per cent of the country's electricity. These reactors were severely damaged in what risk managers call a 'common mode' incident – that is, an event that overwhelmed several safety systems at once.

Known as Fukushima I – or by the Japanese term for the same thing, Fukushima Dai-ichi – the complex consisted of six reactors of an old boiling-water reactor type, three of which were operational at the time of the disaster.

A boiling-water nuclear reactor uses uranium oxide as a fuel in its core. The fuel is 'enriched', which means a small percentage of the uranium is radioactive and gives off neutrons that sustain a controlled chain reaction. This process is called fission, and generates a great deal of heat. Nuclear fuel is not 'highly enriched', however, so fission can never accelerate to the level needed to cause a nuclear explosion.

The heat generated by fission in the core is drawn off by flowing a coolant – in this case water – through the core. The heated water produces steam, which drives a turbine, just like in any ordinary power plant, and the turbine generates electricity. The water is then cooled by a separate, external coolant source (usually also water), before being cycled back to the core in a closed, sealed loop.

It is important to keep the coolant flowing through the core. Drawing off the heat from the nuclear reactor core maintains its operating temperature at about 285 °C. Without coolant, the radioactive decay reaction would drive up the temperature relentlessly. At around 1 200 °C, the alloy rods containing the uranium oxide begin to crack and fail. If this happens, the fuel they contain can escape, which can contaminate the cooling fluid and cause radio-activity to escape from the core. This is what is meant by a 'partial meltdown'.

Control rods are used to control the speed of the fission reaction. They are made of a material that absorbs the neutrons emitted by the uranium decay and, when fully inserted into the core, can shut down the nuclear reaction entirely by absorbing almost all the neutron emissions from the uranium oxide. This is a necessary process when the plant needs to be shut down for maintenance reasons, or to remove spent fuel rods (now consisting of 'depleted' uranium oxide, which can be reprocessed).

However, if the control system fails, or if cooling is lost entirely, the fuel in the rods can heat up uncontrollably. At $2\,800\,°C$, the uranium oxide itself melts, forming a lava-like pool of radioactive material that settles on the bottom of the reactor.

At this point, which is described as 'full meltdown', the most important issue is to contain the molten radioactive core and prevent it from breaching the thick layers of protective material that enclose the reactor, and to try to cool it, while making sure that as little as possible of the cooling liquid, which can itself be contaminated, escapes from the system.

Cooling down the molten core is extraordinarily difficult. Although substances like boric acid can be used to act like control rods to absorb neutrons and slow the reaction down, if emergency cooling systems fail, the process can take many months because the source of the heat is so difficult to reach from outside. It's not like a fire you can just put out by starving it of fuel or oxygen.

A nuclear reactor sports several barriers to prevent radioactive substances from escaping from the reactor and into the surrounding environment. Beyond the design of the fuel rods themselves, the whole reaction takes place in a steel pressure vessel designed to withstand the stress that would occur in the event of an accident. In turn, that vessel, along with the associated plumbing, is housed inside a massive, airtight primary containment vessel made of concrete and steel. A so-called secondary containment structure of concrete surrounds the primary vessel. None of this is visible from the outside, because all of these layers of protection are themselves housed inside the reactor buildings.

The entire structure is designed for one thing and one thing only: to ensure that even in the event of a full core meltdown, the entire disaster can be contained, indefinitely.

The Fukushima reactors are among Japan's oldest, and had been commissioned in intervals beginning in 1971. They had operated without major incident ever since, come earthquake or, indeed, high water. But they were not built for an event as big as the 2011 earthquake.

While the Japanese reactors were not as decrepit and flawed as the infamous Chernobyl, the Ukrainian reactor that melted down in catastrophic fashion in 1986 shortly before the Soviet Union collapsed altogether, there were a few weaknesses in the General Electric design.

In particular, periodic official reports, some as early as 1990, worried about the risks of a simultaneous failure of cooling systems, electricity supply and backup power supplies in regions where major earthquakes and flooding occur. Japanese safety regulators highlighted this risk in a 2004 report, but it was never addressed by the operators, the Tokyo Electric Power Company.

The dangerous scenario became reality on 11 March 2011, and whether it was cost-cutting zeal, honest error or reckless arrogance, it cost the shareholders of the Tokyo Electric Power Company dearly. They lost not only one of the biggest power stations in

Japan, but the entire business as well. It was taken over by the Japanese government, which was the only entity that could handle the massive liability claims and clean-up operation.

As soon as the earthquake struck, several nuclear power plants in the region entered emergency shutdown mode, known as 'scram'. In the cases of Tokai, Higashi Dori, Onagawa and Fukushima II, these shutdowns worked as designed. However, the shutdown at Fukushima I was interrupted by the tsunami forty-six minutes after it began. The control rods, designed to stop the fission reaction by absorbing neutrons, were successfully inserted into the reactor cores, but nuclear decay within the fuel rods themselves continued to generate heat equivalent to about 7 per cent of the heat generated by a fully operational core. Even in full shutdown, maintaining some cooling remains important. However, waves of over fourteen metres hit the complex and penetrated deep into the systems. All power was lost, except for one backup generator, and all instruments and control systems for reactors 1 to 4 went dead. Site infrastructure, including heat sinks designed to draw heat from the cooling fluid, was severely damaged or destroyed.

Batteries kept the emergency cooling going for another eight hours, but when these ran out, so did the ability of the plant to cool itself. Without cooling, it is only a matter of time before the reactor fuel overheats and reaches the failure temperature of the fuel rods.

Over the weeks and months following the accident, a tremendously complicated operation was carried out to cool down the reactors, but the efforts were too little too late. By June, Japan's Nuclear Emergency Response Headquarters released several scenarios of the likely status of the three reactors, but the worst of these raised the deadly spectre: 'meltdown!'

The first reactor was known to have entered meltdown within a day after the tsunami struck.

In their June report, the authorities admitted it was possible that all three reactors had suffered meltdown, although they couldn't say this with absolute certainty. On the upside, attempts to cool down the reactors had met with significant success in the two months after the accident. It wasn't until November, however, that the operators could report that 'cold shutdown' of all the Fukushima reactors had been achieved.

Risks still remain despite the optimistic news towards the end of 2011, and much contaminated water and spent or molten fuel remain on the site. It will be contained there for at least ten years before decommissioning can begin. The final clean-up will take thirty years.

EXAGGERATING THE VERY SERIOUS

The incident at Fukushima was certainly no trivial accident. It was a slow-motion disaster, with plenty for anti-nuclear activists to sink their teeth into.

First, there were hydrogen explosions, which occur when super-heated steam builds up and explodes. These explosions damaged several reactor buildings and caused some radioactive material to be released, though they did not appear to have breached the containment vessels themselves.

To prevent further explosions, steam was periodically vented into the atmosphere. Although this is a controlled process, and the gas is pumped through filters and scrubbers to remove the worst of the contaminants, some radioactive substances will escape along with the steam. They do so in very low concentrations, however, and pose very little actual risk to health or the environment. Thankfully, the weather played along and blew most of the contamination off-shore, where it could decay harmlessly and disperse to little more than background levels.

In the absence of cooling systems, seawater was used to cool the pressure vessels. Not being designed for vast amounts of new water (and corrosive salt water to boot), the waste-water tanks quickly filled, and contaminated water had to be released into the ocean. Cracks in the concrete structure were also a suspected source of contaminated water.

Enter the journalists. Sky News managed to find a 'source' in the person of Holly Williams, a 'Sky News correspondent', who told the television station that radiation in the sea near the plant was now 4 000 times the legal limit.

Of course, the legal limit is pretty near zero. Several thousand times virtually nothing is not a great deal, but Sky's expert source didn't tell us this. She also omitted the fact that the contaminant was iodine-131. This radioactive isotope of iodine is fairly harmful if ingested in significant quantity, but exposure can be easily treated. More importantly, the isotope has a half-life of only eight days. 'Half-life' is the time it takes for half of a given mass of a radioactive isotope to decay into its non-radioactive base element. In the case of iodine-131, halving every eight days means that after thirty-two days, only one-sixteenth of the original radioactive substance will be left.

Moreover, the Pacific Ocean is pretty big. In fact, it contains 622 million cubic kilometres of water. That is 68 trillion times the amount of water estimated to have been discharged from Fukushima. In homeopathic terms, that would amount to a 7C solution, or a ratio of one in ten to the fourteenth power.

That's still not a comprehensible context, so let's try it another way. If you take one millilitre of contaminated water and distribute it equally among a quarter of a million Olympic-sized swimming pools, you'd get the concentration represented by Fukushima's discharge into the Pacific Ocean. True, it will take some years for the contaminated water to spread entirely away from the coast of Japan, but as I put it in a column for the *Daily Maverick* on 4 April 2011, 'a few weeks from now, all that will be left behind is some of the burny stuff your mother used to put on scrapes, albeit in concentrations so low that only a committed homeopath would benefit from bathing in Japanese coastal waters'.[1]

Of course, even being able to measure such radiation is enough to base a headline on if you're writing a sensational article for an environmental organisation or newspaper. For example, an article in the *New York Times* on 31 March 2011 was titled 'Dangerous levels of radioactive isotope found 25 miles from nuclear plant'.[2]

That's true, though the story also admits that the measurement probably only represents a 'hot spot'. In any case, it would have to remain at that level for decades, while you're constantly exposed to it, before you'd notice a measurable increase in cancer risk.

According to the World Nuclear Association, as many as 545 atmospheric nuclear weapons tests have been conducted since World War II. The effects of three nuclear power station accidents, of which only one was truly serious, pales into insignificance compared to the radioactive fallout from these nuclear explosions. And even those didn't add up to a catastrophe that killed millions of innocent little children.

Despite routine warnings from environmentalists that radioactivity lasts practically forever, the background radiation resulting from long-lived substances known to have originated from weapons tests, such as caesium-137, strontium-90 and carbon-14, is more than twenty times lower today than it was in 1963, when atmospheric tests were ended by most countries other than France and China. The annual exposure from the leftovers of all these bombs is equivalent to a single dental X-ray, or one-thousandth of the average annual background radiation. Even at the vast majority of actual test sites (with one exception, in Kazakhstan), the residual annual radiation levels are about a quarter of what you or I would be exposed to anyway during a year.

At Fukushima, sporadic evidence of contamination mounted over the months following the accident. Most instances involved iodine-131, which is harmful for only a short period. Some, however, involved more serious substances, such as caesium-137, with a much longer half-life. In the case of caesium, the half-life is thirty years, so a significant amount of contamination could pose a long-term danger to human health.

The issue was also confused by the changing and sometimes contradictory rules imposed on Japanese citizens. One day, the water was safe to drink, and the next it was not. One day, agricultural products like vegetables and milk were fine, and the next there was a precautionary ban.

However, by June, the International Atomic Energy Agency (IAEA) reported that despite the releases of radioactivity, and the consequent population evacuations from surrounding areas and other emergency measures regarding foodstuffs and water, 'to date no health effects have been reported in any person as a result of radiation exposure from the nuclear accident'.[3] Never mind deaths. No health effects at all had been seen, with the exception of a few of the nuclear plant workers who risked their health trying to contain the crisis. In fact, the IAEA report was broadly positive about the cooperation of the plant's operators and the Japanese government, and about their emergency response procedures. The report did, however, highlight several important lessons to be learned and applied to other nuclear power stations.

DISASTERS AS ECO-MARKETING

A rational appraisal of what went right and how to fix what went wrong was not on the agenda of environmentalists, however. Eleven days into the disaster, the *Guardian* quoted one such activist, Nikki Clark, speaking on behalf of a campaign to prevent a proposed new nuclear power station at Hinkley, on the Somerset coast of England. Clark said: 'We have definitely had more interest since the events in Japan: when we protested over the weekend people really wanted to talk to us about what we were doing, and find out more about the campaign. I think people are realising just how dangerous nuclear really is.'[4]

But it is not really clear what exactly she means by 'dangerous'.

Perhaps she means it in the same way that the environmentalists objecting to a proposed nuclear power plant in the Western Cape, at Bantamsklip, mean it when they say that the new station would 'destroy the livelihoods of virtually everybody dependent on the natural environment of the area'.[5]

Really? Everybody? Like the trouble they have in Cape Town with Koeberg?

The group describes a power transmission line as 'an army of giant pylons and magnetic pollution of unimaginable proportions'. Imagine. A power line. Across the pristine veld!

'Health hazards' are described by simply listing a series of research reports on the effects of various radioactive substances, some of which aren't even found in nuclear power stations, as if those substances are likely to be strewn about the countryside.

Perhaps Clark means it in the way *Carte Blanche* presenter Annika Larsen did when she interviewed Professor Richard Cowling of Nelson Mandela Metropolitan University. Describing the environment around Thyspunt, near Port Elizabeth, which is another proposed nuclear power plant site, he mentions the region's wetlands and shifting dunes.

Quoth Larsen: 'So, the proposed site is actually on shifting sand?' She got the reply she wanted: 'Ja.'[6]

What exactly are they trying to imply here? That a nuclear power station would have no foundation, and would sink into the sands to vanish without a trace? This question on Larsen's part is either staggeringly ill-informed or deliberately obtuse and designed to mislead viewers.

This isn't the only example of exaggeration or outright non sequiturs in the *Carte Blanche* propaganda piece. Try this quote from Saliem Fakir of the World Wide Fund for Nature: 'We've inherited a nuclear sector, an industry, given the apartheid government's involvement in the building of atomic bombs.'

That's a neat bit of rhetoric, associating nuclear power with both atomic bombs and apartheid, without even the hint of a causal connection between them. What is Fakir trying to say? That building another nuclear power station would bring H.F. Verwoerd back from the dead, armed with nuclear warheads?

Or take Wilfred Chivell, of the Dyer Island Conservation Trust: 'If the animals [sharks and whales] would cease to exist, if the animals would cease to come here, obviously you would kill the whole [tourism] industry which brings hundreds of millions to the area and to South Africa as a whole.'

That's true. The thing is, the animals won't cease to exist just because there happens to be a nuclear power station on an isolated spot on a beach somewhere. Contrary to what the Save Bantamsklip Association tries to imply, a nuclear power plant won't be pumping strontium-90 into the ocean to kill everything within a day's sailing.

This kind of anti-nuclear activism is pure alarmist waffle, devoid of any factual basis at all. It doesn't belong on *Carte Blanche*, and it doesn't belong in the *Guardian*.

Perhaps Clark of the Hinkley campaign is of the same school of thought as the renowned 'nuclear expert' Christopher Busby, who immediately after the Japanese disaster went on television to say that Fukushima was already worse than Chernobyl had been.

At Chernobyl, unlike at Fukushima, molten core material was ejected when the entire structure, from the inner pressure vessel out, exploded as a result of hydrogen build-up. That anyone would describe Fukushima as equally serious is odd. That a supposed 'nuclear expert' does so is laughable.

Dr Busby is not just any old crackpot who makes an easy target. He has a Ph.D. in chemical physics, is an active academic and a director of the environmental consultancy Green Audit, was the spokesperson on science and technology for the Green Party of England and Wales, and has served in several advisory roles to the British and European governments. Nonetheless, he is moderately famous for a number of reasons that inspire little confidence in his characterisation of the Fukushima disaster. For example, he has a curious theory that radiation in very small doses is actually worse for you than higher levels of radiation exposure.

His theory, strongly held and vigorously sold as a self-published manuscript on his website for £25 a copy, was roundly rejected by a committee that was asked in 2001 by the then British minister of the environment, Michael Meacher, to examine the risks of radioactive substances that people ingest or inhale, for the most part naturally. The irony? Busby himself was one of three active environmental anti-nuclear campaigners on the committee that evaluated and rejected several major aspects of his own work. He objected, of course, but was overruled.

Besides his rejected scientific theories, the fellow also flogs a range of mineral supplements, which he claims will prevent heart attacks in children resulting from radiation exposure one-sixtieth as strong as the radioactive dose they get from, say, a banana. In a report that quotes a certain Busby quite a lot, Busby explains how this all works, but adds that the supplements are not very effective because he's not allowed to put stable caesium in them. Still, he markets them under the label 'Busby Laboratories, Formula 1, Christopher Busby Foundation for the Children of Fukushima'.[7]

This environmentalist charlatan was all over the media in the days following the tsunami, warning of the dangers of 'concentrations of plutonium and caesium and iodine' that he claimed are coming down even hundreds of kilometres from the stricken plant. He told the BBC three days after the tsunami that a meltdown could result in a 'nuclear explosion'.[8]

Such blatant nonsense ought to immediately disqualify him from ever speaking on the subject of nuclear power again.

Nuclear reactors do not use weapons-grade substances that can cause 'nuclear explosions'. They don't use any other substances that can cause 'nuclear explosions' either. Such explosions are a scientific impossibility.

The explosions that did occur at Fukushima were simple hydrogen explosions resulting from superheated steam, and they damaged only the external buildings, not the containment vessel that is designed to keep nuclear fuel from escaping. Even the massive explosion at Chernobyl, which did rupture the containment vessel and threw nasty stuff sky-high, was not a nuclear explosion. A layperson might be excused for confusing an ordinary combustive explosion with a nuclear chain reaction, but from a so-called nuclear expert, the phrase is a flat-out lie.

As for Busby's claims about plutonium, let's not quibble about the fact that a 'nuclear expert' might be expected to distinguish radioactive isotopes from their basic elements and specify the isotopes that he was concerned about. It is true that some plutonium isotopes were found near Fukushima. As Reuters reported breathlessly in an article that also exaggerated the number of dead or missing by 50 per cent: 'Plutonium found in soil at the Fukushima nuclear complex heightened alarm yesterday about Japan's battle to contain the world's worst atomic crisis in 25 years as pressure mounted on the prime minister to widen an evacuation zone around the plant ... A by-product of atomic reactions used in nuclear weapons, plutonium is highly carcinogenic and one of the most dangerous substances on the planet, experts say.'[9]

If significant quantities of the stuff had been found, this would indeed be alarming. Not least because it would imply that Fukushima was a super-secret nuclear weapons factory, although the inconvenient absence of evidence did not stop lunatic-fringe conspiracy theorists from making exactly this claim on message boards such as that run by Busby. But the same article, albeit only in its very last paragraph, concedes that the plutonium levels were so small that they were not harmful. Fears that the samples could indicate a breach of the containment vessel were justified, especially since the samples contained a particular isotope that was not likely to have come from a nuclear explosion but could be produced by a reactor such as the ones at Fukushima.

However, the majority of the plutonium found in the soil near Fukushima appears to have been residue from atomic testing in the Pacific many years before. Either way, as MIT scientists quickly pointed out, despite its radioactivity and chemical toxicity, 'no human has ever died from acute uptake of plutonium'.[10]

COULD THIS HAPPEN TO KOEBERG?

The South African public had already been primed to distrust nuclear power by media 'exposés' such as the *Carte Blanche* programme mentioned above, and another screened in November 2007 titled 'Uranium Road'. This is a documentary made by Teaching Screens Productions, and, much like the anti-fracking screed *Gasland*, is a firm favourite on the independent film circuits on campuses and in art-house cinemas.

'Uranium Road', whose transcript is (perhaps for copyright reasons) now curiously missing from the *Carte Blanche* archives, ominously referred to the powerful lobbies that support nuclear energy, but did not bother to name them. It did, however, extensively quote one David Fig, who was described innocuously as an independent researcher. As it turns out, Fig was not only the chairman of a vocal environmentalist outfit named Biowatch South Africa, but he had a book out. The title? *Uranium Road.*

The classic environmentalist technique of rolling out evocative images and vague associations came thick and fast. As Darren Olivier, then a contributor to the award-winning blog *Commentary South Africa*, told me, this was not investigation, this was advocacy.

The apartheid-era nuclear weapons programme was mentioned on the *Carte Blanche* show, just to refresh the association. An idyllic picture of a farmer tending his cows with the Koeberg nuclear power station in the background was accompanied by an ominous voiceover describing the uptake of heavy-duty nuclear waste into the environment, as if it just gets dumped in the local landfill. Heartrending images of the victims of Hiroshima and Nagasaki were shown, as if nuclear power has anything in common with atomic weaponry.

Or does it? The film aired on *Carte Blanche* (as well as Fig's book) darkly hinted that the Pebble Bed Modular Reactor project, a now-shelved attempt to build an inexpensive, safe, modular nuclear power station, might be a cover for an ANC plot to build nuclear weapons. They must think the terrorist plotters in the ANC are too stupid to just use the Koeberg power station and the nuclear research plant at Pelindaba in Gauteng for their nefarious plots to nuke, well, Lesotho maybe.

The Nuclear Industry Association of South Africa did not cover itself with glory in response to this show. It cack-handedly threatened to haul *Carte Blanche* before the Broadcasting Complaints Commission. A powerful lobby that masterfully manipulates both government and the public, as *Carte Blanche* would have it, might have thought it smarter to coolly disseminate a document or video refuting the many exaggerations and errors in the show. It wouldn't have been hard to do.

The intrepid crew at *Carte Blanche*, who once devoted an entire show to an ex-cop with a magic woo-woo invention for locating missing people, also ran a story about another scare: that Koeberg is located near a fault line and could therefore be just as vulnerable as Fukushima.

'Koeberg sitting alongside fault line,' echoed the *Cape Times* just days after the Fukushima disaster.[11]

Factually, that is correct. The Koeberg plant is indeed located near a fairly minor fault line known as the Milnerton Fault. And it is also true that this fault has produced earthquakes in the past. But the rest of the story is a loose amalgamation of facts and allusions, culminating in the use of the term 'liquidification' instead of 'liquefaction' to describe what happens to certain sedimentary soils during an earthquake.

The two worst earthquakes in South Africa's history were both estimated at 6.3 on the moment-magnitude scale. One of them, in 1809, did occur on the Milnerton Fault. Koeberg was designed to survive an earthquake of magnitude 7, which is five times as strong (the scale is logarithmic, not linear) as the worst ever documented in South Africa's history. An insurance company that worked on the 2010 World Cup bid calculated actuarial scenarios for South Africa, and concluded that quakes of magnitude 6 or higher happen once every 300 years. Even its very worst-case scenario fell a little below what Koeberg was designed to handle. By comparison, the earthquake that hit Fukushima was two-and-a-half times stronger than what the plant was designed for.

In the US, similarly panicked warnings followed a magnitude 5.8 earthquake in Virginia in September 2011. By design, this prompted an automatic shutdown of the North Anna nuclear plant, situated about eleven miles from the epicentre. The shutdown was entirely precautionary. Although the quake was the most severe east of the Rocky Mountains since the late nineteenth century, it was well within the design capacity of nuclear power plants in the US, and therefore no cause for alarm.

PUBLIC OPINION VERSUS REALITY

The environmental organisation Friends of the Earth commissioned an opinion poll in Britain in the days following the Fukushima meltdown, and promptly got the media to give publicity to its findings. According to this survey, support for new nuclear power stations fell by 12 per cent since Fukushima, and outright opposition grew by 9 per cent, the *Guardian* reported on 22 May.

This was, technically speaking, true, and thanks in no small measure to the alarmism of environmentalists. However, what the newspaper reports of this survey did not mention was that even after this change, more people expressed support for nuclear power than opposed it. Instead of 'Japan nuclear crisis puts UK public off new power stations', a more accurate headline would have been 'Despite Japan nuclear crisis, UK public still favours new power stations'.[12]

Moreover, the change was estimated by comparing two surveys conducted by two different organisations using two different methodologies. In textbooks on market research, this kind of comparison between non-comparable studies is a great big no-no, since the cause for any disparity could very well have something to do with how the survey was conducted or analysed. It should have raised red flags. Journalists ought to know better than to parrot the press releases of environmental groups pushing their agendas, even if they have harmless-sounding names like Friends of the Earth. Such groups are no less capable of exaggeration and spin than the corporate giants whose press releases journalists routinely shun as a matter of professional pride.

In truth, few journalists bother to question received wisdom, or endeavour to correct misconceptions about nuclear power and accidents. Sensationalism sells, research is hard and conclusions that debunk the intellectual elite's treasured environmental myths are unpopular.

SPRING HAS SPRUNG, THE GRASS IS GREEN

One of these persistent myths, another version of which we'll come across again in the next chapter on oil spills, was dispelled by Catherine Brahic, writing in the *New Scientist*. 'What does a coral reef look like 50 years after being nuked? Not so bad, it seems,' she wrote.[13]

Only a massive crater is left of the three islands that once were part of Bikini Atoll, the site of above-ground nuclear weapons testing in the 1950s. Yet according to Brahic's article, Zoe Richards, of the ARC Centre of Excellence for Coral Reef Studies in Australia, says it is thriving, with 183 species of coral having been counted, some of them growing 'like trees', as much as eight metres high. Richards and her colleagues confirm that the ambient radiation in the area is low, but people have been scared off by the post–World War II fear of nuclear weapons, and by the post–Cold War fear of nuclear anything.

This isolation has helped the recovery of the local ecosystem, and the researchers estimate that 65 per cent of the biodiversity that existed in the region before it was blighted by weapons testing has since bounced back.

The biodiversity of the Chernobyl exclusion zone is a more controversial question. Several scientists, such as Viktor Dolin, of the Ukrainian National Academy of Sciences in Kiev, and Sergii Gashchak, a researcher at the Chornobyl Center [sic] in Ukraine, said that the area is a biodiversity hot spot, with some 100 threatened and endangered species thriving in the deserted area. Although there are animals with radioactivity-induced mutations, they say, these appear to die off young, leaving healthy specimens to mature and procreate.

At a conference in Johannesburg in 2009, Hennie Eksteen, a vermiculturist of international repute – in circles where worm experts are famous, at least – gave a talk titled 'How earthworms save the world'. Eksteen explained many mildly astonishing features of these humble creatures, including this one: 'They can also encapsulate toxic heavy metals before excreting them, which leaves the metals isolated in a hard shell. This is what happened after Chernobyl, and why the local environment looks so good now.'[14]

However, a recent counter-study suggests a decline in biodiversity in the exclusion zone compared with non-contaminated regions. Reporting on this study, one green blogger said that we probably all feel 'NIMBY' about nuclear power stations, meaning 'not in my back yard'.

Ironically, this is also a misconception. If you had to choose, you'd be better off with a nuclear plant next door than with a coal-fired plant. What you don't hear quoted very often is the title of a 2007 *Scientific American* article: 'Coal ash is more radioactive than nuclear waste'. Writes Mara Hvistendal about the notion that coal might pollute, but isn't as dangerous as nuclear power:

Over the past few decades, however, a series of studies has called these stereotypes into question. Among the surprising conclusions: the waste produced by coal plants is actually more radioactive than that generated by their nuclear counterparts. In fact, the fly ash emitted by a power plant – a by-product from burning coal for electricity – carries into the surrounding environment 100 times more radiation than a nuclear power plant producing the same amount of energy.[15]

Overall, the report on which she bases her article says that people living around a coal-fired power station are exposed to between two and four times as much radiation as those living near a nuclear power plant. Not to mention, of course, all the other pollutants from coal-fired stations that nuclear plants don't generate.

However, atypically for an article warning about the dangers of radiation, this story does provide some context about the true scale of the radioactive danger. The natural background radiation from rocks, food and other ambient sources, it reports, is still 200 times higher than the likely annual exposure from even a coal-fired power station.

LET'S TALK ABOUT DEATH

The Three Mile Island nuclear power plant in Pennsylvania was the site of a partial meltdown in 1979, which prompted the first fears about nuclear power stations, as opposed to nuclear weapons.

However, more than thirty years later, in the days after Fukushima, Dr Ian Fells, emeritus professor of energy conversion at the University of Newcastle upon Tyne and former chairman of the New and Renewable Energy Centre at Blyth, Northumberland, in England, had to remind the BBC's other guest, Christopher Busby, that there was no dangerous level of radiation exposure to the surrounding population, 'despite what these days a lot of people think'.[16]

Extraordinary numbers are also bandied about with regard to Chernobyl, which really was a major disaster and really did kill people – unlike either Three Mile Island or Fukushima. For example, another so-called expert quoted in both the international and South African media, Natalia Mironova, agreed with Dr Busby's assessment that Fukushima was worse than Chernobyl. In describing Chernobyl, however, how many people did she think died?

She didn't mention that of the 600 workers on the Chernobyl site when the misguided experiment to test the badly designed reactor's emergency cooling systems went catastrophically wrong, 134 received high doses of radiation and suffered from radiation sickness. Of these, twenty-eight died in the first three months, and a further nineteen died in the next eighteen years from causes that might be linked to radiation.

She didn't refer to the massive number of recovery operation workers – 530 000 of them – all of whom on average received enough exposure to show a statistically noticeable increase in cancer risk. Nor was she satisfied by counting the number of local residents who were exposed. Children were particularly vulnerable to thyroid cancer as a result of milk contaminated with iodine-131 in the days immediately after the disaster. Unlike at Fukushima, precautionary measures that drew attention to the government's failures were not popular in the Soviet Union.

A 2005 report by the WHO and the IAEA estimates a total number of excess cancer deaths worldwide – the number of deaths over the background rate – as a consequence of Chernobyl of about 9 000. Mironova did not quote this figure.

A more recent study by the UN's Scientific Committee on the Effects of Atomic Radiation finds the total to be about 27 000. The report for the British government by the Committee Examining Radiation Risks of Internal Emitters, on which Busby sat and which assessed these estimates, broadly endorses estimates in this order of magnitude. The phrase they most often use is 'slight increase'. Mironova didn't use this finding either.

Greenpeace, not known for its conservative estimates, said in a 2006 report that Chernobyl caused almost 93 000 fatal cancers. Not good enough for Ms Mironova.

The editors of that report, by the time they got to writing a book about it, revised that number to 980 000, making it ten times worse. Veteran anti-nuclear campaigner Helen Caldicott, who runs the Helen Caldicott Foundation for a Nuclear-Free Planet, cited this number in an interview with CNN shortly after Fukushima – but not Mironova.

In 2011, one of the very same editors, Alexey Yablokov, ventured to estimate 1.4 million excess deaths. Busby himself, in his own review of the numbers – which, as we've seen, disagree quite remarkably from those of even the committee on which he sat – finds this quite a reasonable number. Mironova would disagree.

Natalia Mironova chose none of the above. To quote the article: 'Mironova said Chernobyl would likely impact the health of 600 million people around the world over the long-term, or nearly nine times more than were killed in World Wars I and II.'[17]

One should grant that 'impacting the health' is not quite the same as killing, but when we're talking about cancer rates, and the WHO reports any claims of even just tens of thousands of deaths to be grossly exaggerated, such a big total is scarcely credible.

Still, one has to concede that 600 million is a terribly scary number. Well done to Mironova for making it up.

It is worth contrasting this technique of inflating death rates as a consequence of nuclear accidents with what newspapers did with the almost 20 000 people that did not die at Fukushima, but did die as a result of the earthquake and tsunami. In many stories, even shortly after the event, these ever-climbing statistics were simply relegated to the end of the story, below the more important headlines scaring readers senseless with fear over leaking radioactive waste water or contaminated food.

A little factoid has always been stuck in my head: energy derived from coal is responsible for thirty mining deaths per day, making it by far the most deadly source of energy. The problem with this bit of trivia is that it is vulnerable to the claim that there are many more coal mines in the world than, for example, uranium mines. This is true. A better measure would calculate deaths per energy source as a consequence of both operational accidents and pollution, as a function of power produced. *The Next Big Future* blog has done exactly that, sourcing data from various agencies to produce a comparison. The startling result, that nuclear is by far the safest form of energy, can be seen in Table 5.1.

Table 5.1
Deaths per terawatt-hour (TWh), by energy source

Energy source	Death rate (deaths per TWh)
Coal (China only)	278
Coal (world)	161
Oil	36
Coal (USA only)	15
Biofuel/biomass	12
Peat	12
Natural gas	4
Hydro (world including 170 000 Banqiao Dam deaths)	1.4

Table 5.1
Deaths per terawatt-hour (TWh), by energy source

Energy source	Death rate (deaths per TWh)
Solar (rooftop)	0.44
Wind	0.15
Hydro (Europe)	0.1
Nuclear (including Chernobyl and Fukushima)	0.04

Source: 'Deaths per TWh, by energy source', The Next Big Future, 13 March 2011, http://nextbigfuture.com/2011/03/deaths-per-twh-by-energy-source.html.

HOW TO AVOID THE FACTS

The pattern in all the exaggerated stories, whether by environmentalists or by the newspapers that quote them, is the same. Always describe the worst possible effects of a substance. Describe the horrible symptoms of acute radiation poisoning or a high dose of toxic chemicals, even when you're talking about minimal exposure.

If you really can't avoid references to inconvenient facts that show the situation isn't quite as bad as you'd like journalists or readers to think, bury those facts towards the end of the story. For example, describe 'radioactive milk' or 'contaminated fish', even though the measurements are barely detectable and extremely unlikely to pose any serious risk. Such trivial details are for paragraph 14, where hopefully they'll get cut for space reasons.

Make sure to confuse nuclear power with the awfulness of nuclear weapons. Caldicott does this well, for example, when she tells CNN that spent-fuel pools 'contain two to 10 times more radiation than in the reactor core, which itself contains as much long-lived radiation as 1,000 Hiroshima bombs'.[18] That is factually correct, of course, but spent-fuel pools aren't exploded in mid-air over densely populated cities.

Always try to use adjectives creatively. A *New York Times* article on the evacuation around Fukushima, for example, was headlined 'Japan quietly evacuating a wider radius from reactors'. The article began this way: 'Japanese officials began quietly encouraging people to evacuate a larger swath of territory around the Fukushima Daiichi nuclear plant on Friday, a sign that they hold little hope that the crippled facility will soon be brought under control.'[19]

What they mean by 'quietly' is that a government minister gave a press conference that was broadcast live on television to ask people to leave. The announcement couldn't possibly have been made more publicly or more loudly.

When the underhanded use of adjectives was ridiculed by bloggers, the *New York Times* 'quietly' changed the article – without even the notification that usually accompanies a correction – to remove all references to 'quietly'.[20]

Once your adjectives are suitably scary or sinister, exploit the fact that 'normal' or 'legal' limits are extremely low. For example, the exposure level known to cause detectable statistical cancer risk is eight times higher than the legal radiation-exposure limit for US nuclear workers. It is 400 times higher than the annual exposure limit for ordinary members of the public. Given such remarkably low limits, it is easy to create headlines that involve scary, big numbers. Your exposure can be fifty times higher than normal and still be within the ordinary occupational health and safety standards for nuclear power plant workers. When the norm is virtually zero, it's pretty easy to get to 'many times worse'.

You see, facts are not always material to environmentalists. I first wrote a column warning against overreacting to Fukushima on the very weekend of the disaster, before its full scope was evident. It was published early in the morning of 15 March 2011. In it, I used an argument I've used often before and since – namely, that although aircraft accidents are horrible and make for lurid headlines, they offer an opportunity to learn and to improve safety standards, not to stop flying.

The first reader to object accused me not only of poor timing (what better timing could there be?), but also of plagiarising an article in the *Economist,* which used a similar line of reasoning. The *Economist*'s article appeared on the same day, a day after my column was submitted to my editor.

Minor issues such as the logical constraints imposed by the physics of the space-time continuum, which prevent me from copying from the future, have never stopped environmentalist critics from making grave accusations to try to discredit those with whom they disagree.

THE SUM OF ALL FEARS

Not all environmentalists are opposed to nuclear power. James Lovelock, the father of the Gaia hypothesis – about which I'll have more to say in a later chapter – is a famous proponent of the technology. And at least one prominent environmentalist, George Monbiot, has publicly changed his position on nuclear power as a consequence of the Fukushima disaster. Unlike Germany and Japan, he hasn't turned against it. Surprisingly, he has become a supporter.

In a widely disseminated (and viciously criticised) article for the *Guardian*, George Monbiot explains:

As a result of the disaster at Fukushima, I am no longer nuclear-neutral. I now support the technology. A crappy old plant with inadequate safety features was hit by a monster earthquake

> and a vast tsunami. The electricity supply failed, knocking out the cooling system. The reactors began to explode and melt down. The disaster exposed a familiar legacy of poor design and corner-cutting. Yet, as far as we know, no one has yet received a lethal dose of radiation. Some greens have wildly exaggerated the dangers of radioactive pollution.[21]

He cites many of the points made in this chapter, and a few more. He describes the likely consequences of the exaggerated fears, and says they are likely more serious than the fears themselves: either we resort to the kind of small-scale energy projects that ruined the landscape and ecology of Europe during the industrial revolution and rendered the people of the eighteenth century much poorer than even the poor of today, or we turn away from nuclear towards what Monbiot considers an even worse alternative, fossil fuels.

This is especially true for developing countries like South Africa. In the midst of a grave energy crisis, which will take years to resolve even with the cheapest, fastest, dirtiest coal-fired projects in the works, environmentalists like those in the Save Bantamsklip Association are screaming blue murder about … power lines.

When industrial capacity is so low that a quarter of the population is unemployed, and the state-owned power utility keeps calling on industrial users to cut electricity use, the greens are raising the alarm about, almost literally, a drop of iodine-contaminated water in the ocean.

When mining output in South Africa is shrinking, in the face of growing output everywhere else in the world, and when one of the reasons is the lack of reliable power supply, concerned residents, academics and journalists peddle tall tales about 'liquidification' and 'shifting sands'.

If poverty, hunger and unemployment are the true crises of

today, while incidents such as the Fukushima meltdown are not only rare but also much less damaging when measured in blood and treasure than the natural disaster that precipitated it, one has to question the priorities of environmentalists, tenured professors and the mainstream media.

One may disagree with a lot that Monbiot says on other subjects, but he's quite correct in his conclusion about nuclear power:

> Yes, I still loathe the liars who run the nuclear industry. Yes, I would prefer to see the entire sector shut down, if there were harmless alternatives. But there are no ideal solutions. Every energy technology carries a cost; so does the absence of energy technologies. Atomic energy has just been subjected to one of the harshest of possible tests, and the impact on people and the planet has been small. The crisis at Fukushima has converted me to the cause of nuclear power.

Christopher Busby, in typical exaggerated style, called Monbiot 'criminally irresponsible' for saying so.[22] This, all by itself, is proof positive that Monbiot got this call right.

6

The Caribbean
Reports of My Death Were Greatly Exaggerated

'A HOLE IN THE WORLD'

It was 'not just an engineering accident', wrote the ever-hyperbolic Naomi Klein in the *Guardian*, but a 'violent wound inflicted on the Earth itself', a 'violent wound in a living organism'.[1] The explosion of Deepwater Horizon, the ultra-deep-water oil-drilling rig that suffered a well blowout in the Gulf of Mexico on 20 April 2010, was 'a hole in the world', read the headline.

The rig, owned and operated by a specialist contractor for BP, was working on a site known as the Macondo field, about 70 kilometres off the coast of Louisiana, near New Orleans. The sophisticated rig, worth over half a billion dollars, was completed in 2001. Until its untimely demise, it had enjoyed a relatively blemish-free safety record, aside from a handful of citations in the first couple of years of its working life. In 2008 the rig won an award as an industry model for safety, and in 2009 it broke the record for the deepest ever deep-water oil well, going over 10 kilometres deep in waters that would sink Table Mountain.

Ironically, managers of both BP and the contractor, Transocean, were aboard the rig only hours before the blowout to celebrate seven years without serious accidents. Deepwater Horizon sank thirty-six hours later.

The cause of the accident was a failed cementing job by another subcontractor, and it was soon discovered that the well was leaking. Standard safety measures to contain such a well blowout failed, and for eighty-seven days following the explosion, thousands of barrels of oil per day leaked into the sea. By the time a solution was found and the leak was plugged, four million barrels of black gold had been lost to the Gulf of Mexico. A massive clean-up operation had begun almost immediately after the accident, and BP soon realised it was on the hook for billions in damages.

Company representatives were facing questions from angry local residents at a town-hall meeting when Klein swooped in with a documentary team from Al Jazeera. Klein attained fame with a manifesto against capitalism, consumerism, corporatism and globalisation, entitled *No Logo*, and has been hailed by the *New Yorker* as 'the most visible and influential figure on the American left'.[2]

Not content with challenging the failings of rig workers, oil company executives and safety regulators, all of whom shared a measure of responsibility for the accident, Klein even took a swipe at a local school football team known as the Oilers. They were 'unfortunately named', in her opinion, as if they had no right to respect the hard and dangerous work of many of their parents, who help to supply the world with the energy it needs to fuel productivity and prosperity.

While even the children got caught in the scattergun of her sanctimonious condemnation, nowhere in a 4500-word feature-length article could she spare even a single line to acknowledge the eleven oil rig workers who died in the explosion. Their names were Jason Anderson, Aaron Dale Burkeen, Donald Clark, Stephen Curtis, Gordon Jones, Roy Wyatt Kemp, Karl Dale Kleppinger, Blair Manuel,

Dewey Revette, Shane Roshto and Adam Weise. Eleven safety helmets cast in bronze serve as a memorial, and children placed white roses around a model of the rig in remembrance of the dead.

According to the Pew Research Center, the media coverage of the Gulf oil spill was unusual in that it continued to feature prominently for more than three months after the accident. By contrast, the Virginia Tech campus massacre in which thirty-three people died, a collapsed bridge in Minneapolis that killed thirteen people, and a mine accident in West Virginia in which twenty-nine people lost their lives had all vanished from the headlines two weeks after the event.

Not so the Gulf oil spill. It dominated news coverage for 100 days after the accident, receiving more than twice as much space as the next most important topic – the economy. It was a complex and ongoing story, told with dazzling infographics and dramatic footage of oil-covered birds.

Let's be clear. The explosion that led to the sinking of the rig was certainly a catastrophe. In addition to the tragic loss of human life and seventeen others injured, shareholders took a massive blow, losing £50 billion of their capital in just two months as the share price of BP plummeted. The company set aside an initial £20 billion to pay compensation claims, but warned that this did not include fines and penalties related to the accident, nor would it represent a cap on the company's potential liabilities. Criminal charges against engineers alleged to have misled safety inspectors have not been ruled out.

To be fair to Ms Klein and the environmental lobby, the accident had a significant impact on the Gulf of Mexico. Not surprisingly, an oil spill on this scale does indeed pose a threat to marine life, and many of the anecdotal tales of polluted coastal areas, dying birds and contaminated shellfish were perfectly true.

Was it even possible to exaggerate such a large-scale disaster? You bet it was.

FOR DHARMA'S SAKE, PANIC!

Among the thousands of sober and factual articles describing the complexities of staunching a leaking deep-water oil well, the logistics of widespread clean-up operations, the implications for the environment, and the details of the political, corporate, legal and financial fallout, there was a great deal of shrill hysteria. Some commentators went as far as predicting the death of the entire Caribbean Sea.

This makes the Deepwater Horizon story an interesting case study of environmental exaggeration. Even when the news is terrible, green activists, scientists and journalists not only are able to blow it out of proportion, but feel the need to do so.

The boss of BP, Tony Hayward, thanks in part to his own attempts to play down the seriousness of the accident and a few frankly naive lines he delivered off the cuff, had become 'the most hated and most clueless man in America'.[3] For a CEO who once aimed to pull BP back from the dazzling risk-taking trajectory it had taken in the economic heydays of a decade ago and return it to a sense of conservative safety, this must have hurt.

Six weeks after the event, *Planet Green*, a Discovery Channel website, described the accident as follows: 'An epic environmental tragedy has been unfolding recently, with things going from bad to worse to terrible (and it's not over yet).'[4] Now if you run out of superlatives before a disaster is even over, you may want to revisit your style guide's admonitions against the overuse of adjectives.

Another website, *Treehugger*, also owned by Discovery, referred to a 'Katrina of smell' that is 'attacking New Orleans'.[5] 'BP's underwater gusher in the Gulf has our collective eco-anxiety approaching

11,' it added. It would 'coat hundreds of miles of beaches, endangered seabirds, oyster beds, and protective barrier wetlands'. Fishermen would 'loose [*sic*] their way of life'.[6] Its effects 'will have an impact on wildlife – and the people whose livelihoods depend on it – for generations to come'.[7]

Fears were raised that the oil spill would enter the Gulf of Mexico's main sea current, and if it did so you could 'bet the farm', according to Tony Sturges, professor emeritus in oceanography at Florida State University,[8] that it would go around the corner of Florida into the Atlantic Ocean's Gulf Stream. The supposedly sober and respectable National Center for Atmospheric Research (NCAR), which is sponsored by the US government's National Science Foundation, produced an ominous animation illustrating the possible dispersal pattern of the oil slick, which spread like wildfire across the internet, fuelling the hysteria.

'As oil continues to surge into the Gulf of Mexico from the site of the Deepwater Horizon rig accident,' a breathless *National Geographic* piece began, 'experts warn that the Gulf's powerful Loop Current could whip millions of gallons of oil around Florida's peninsula and north to East Coast beaches.'[9] The oil would not have weathered enough to 'lose its noxious properties' by then, according to the scientists, weather watchers and environmentalists interviewed for the article, and it would end up causing death and destruction in the mangrove swamps, where endangered life forms spawn.

'Where will the Deepwater Horizon oil end up?' asked *Scientific American*. 'The short answer is everywhere – the sea surface, deep waters, the Gulf Coast, in deepwater corals and even as far as the Arctic.'[10]

Several of the more sensationalist newspapers warned of the death of 'an entire sea', and others posted graphics of the 'great ocean conveyor belt', the system of currents that spans the world's oceans, adding, 'Here's a visual of where it can go from there. I'll … let you contemplate this image from The New England Aquarium on your own …'[11]

Jill Schneiderman, a professor of earth science at Vassar College in New York, dubbed the site 'Pandora's well'. To those who 'care about the Earth, or … care about dharma', she wrote that she became a geologist as an 'act of devotion', and considers accidents such as Deepwater Horizon to be among 'the disastrous upshots of our inane technological "achievements" '.[12]

Note the inverted commas around 'achievements'. The use of such scare quotes is a rhetorical technique commonly used to express scepticism or scorn. Schneiderman does not believe that humanity's achievements are really achievements, but that they are merely so called. In fact, she explicitly calls them 'inane'.

Now I don't know about you, but I'm pretty impressed with our 'inane' achievements. We can instantly communicate with anyone else on the planet. We can travel rapidly and in comfort even across vast oceans, and brag about it by complaining about the food served at 30 000 feet. We can detect diseases using complex machines and cure them using sophisticated chemicals, genetic engineering or laser surgery. We build skyscrapers and bridges of gigantic proportions. We can print architectural models, product prototypes and even prosthetic teeth using 3D printers. We feed, for the most part, seven billion people, using little more arable land than we needed when the human population was only a billion strong.

All this might be rather silly to some people – and I sure hope they're consistent and do the washing by hand on Mondays – but I'll bet they aren't quite so whiny when they need to be airlifted to hospital in a helicopter because all that back-breaking labour caused a heart attack.

When one dismisses the rise of human civilisation as something to put in scare quotes, while respectfully deferring to a motley assortment of ancient religious myths, is it any wonder that Schneiderman would describe an accident in near-apocalyptic terms?

A Google engineer created a nifty mapping tool that lets users compare the size of the Deepwater Horizon oil slick to the size of their city. This is, of course, a perfectly meaningless comparison. An oil slick is only a few microns thick, so its area is vast in comparison to its volume. Its impact, however, has little to do with its surface extent. It depends on how much of the oil, by volume, washes onto shores, over sensitive reefs or into productive fishing areas.

But the mapping gimmick was promptly picked up by websites collecting resources aimed at teachers, so that they could teach children in a way that wouldn't bore them to death but would scare them silly. It didn't tell them that there are 24000 urban areas on the planet, covering a total of between 1 and 3 per cent of the earth's surface. So if 'your city' is near average size, it is likely to occupy about one millionth of the planet (0.000125 per cent, to be precise).

The point here is one that is made throughout this book: it is important to see matters in context. The Google comparison is exactly the wrong kind of context for a sober appreciation of the magnitude of the Deepwater Horizon disaster. It's about as useful as scale illustrations that reference the distance between the earth and the moon. Does anyone – even astronauts who have actually been there – really have a good mental picture of how near or far the moon is from the earth in relation to, say, the thickness of a stack of paper or a row of pencils laid end to end?

Such comparisons exist to make children go 'Wow!' at baffling numbers. They are designed to exaggerate. They don't make big numbers any less baffling, and adults shouldn't fall for such trickery.

WHY HAS IT GONE SO QUIET?

As it turns out, the impact of the Deepwater Horizon spill was considerably less severe than anticipated. The official report on the incident runs to almost 400 pages and is no stranger to environmental rhetoric. In addition to the usual well-worn phrases of disaster, it includes heart-rending people-profiles to 'humanise' the accident, and at one point describes an oil spill as an 'insult'.

Yet it has the following to say about the NCAR's alarming animation and the warnings of *Scientific American* and *National Geographic*: 'a large circulating eddy kept oil from riding the Loop Current toward the Florida Keys'.[13] That is, contrary to the flashy government-sponsored Photoshop job circulating on YouTube, the oil didn't even reach the Gulf's own main current to pollute the Florida Keys, let alone reach the Atlantic Gulf Stream to head for North Carolina, Scandinavia and the Arctic.

The report goes on to describe at length how difficult it is to assess the damage to the ecosystem and the economy, in part because no baseline data exist from before the accident with which to compare the situation after the accident. What the report had no difficulty presenting, however, was the 26 per cent increase in clinical depression, and the double-digit increases in reported feelings of stress, sadness and worry, among people living in the US states along the Gulf coast.

A comprehensive wildlife collection programme was undertaken in the wake of the disaster, and it regularly reported data until November 2010, just over six months after the accident. It examined animals that were collected from the region and brought to authorities for assessment. It found many animals that had died, some of which had been exposed to oil.

A hundred dead dolphins were discovered in the area in the six months following the incident, but remarkably only four of them were visibly oiled. Well over a thousand turtles were collected dead or alive, of which more than 40 per cent were visibly oiled. However, of the half of the turtles that had died, the number that were visibly oiled was less than 3 per cent. These are startling discrepancies.

The marine bird numbers are also revealing. The total number collected, dead or alive, exceeded 8 000. Of these, 53 per cent were visibly oiled. Three-quarters of the collected birds were dead, and of these, far fewer (37 per cent) were oiled. The discrepancy becomes clear in light of the fact that every single one of the more than 2 000 live birds collected was visibly oiled. Researchers apparently saw no need to catalogue unaffected animals to provide context for the number of affected animals, unless they were dead, in which case you'd get a bigger number by just listing all of them. Yet, even if all the dead animals had died as a result of exposure to oil in the water, you still need some idea of how big the total population is. If there were 10 000 birds before the accident, it's a catastrophe. If there were a million, it's small potatoes. This kind of systematic bias in the collection of animals makes the resulting numbers entirely useless.

The report on the collection programme does note, appropriately, that the numbers do not reflect the cause of death, that not all the deaths are necessarily related to the Deepwater Horizon impact area, and that the unusually high number of researchers combing the area makes it likely that a higher than normal number of injured or dead animals would have been recovered. However, by the time the environmentalists got hold of the numbers, such disclaimers had long fallen by the wayside. 'More than 8,000 birds, sea turtles, and marine mammals were found injured or dead in the six months after the spill,' says the National Wildlife Federation in its call for help.[14] It does the same for dolphins and turtles, describing every single collected animal as a victim of the accident.

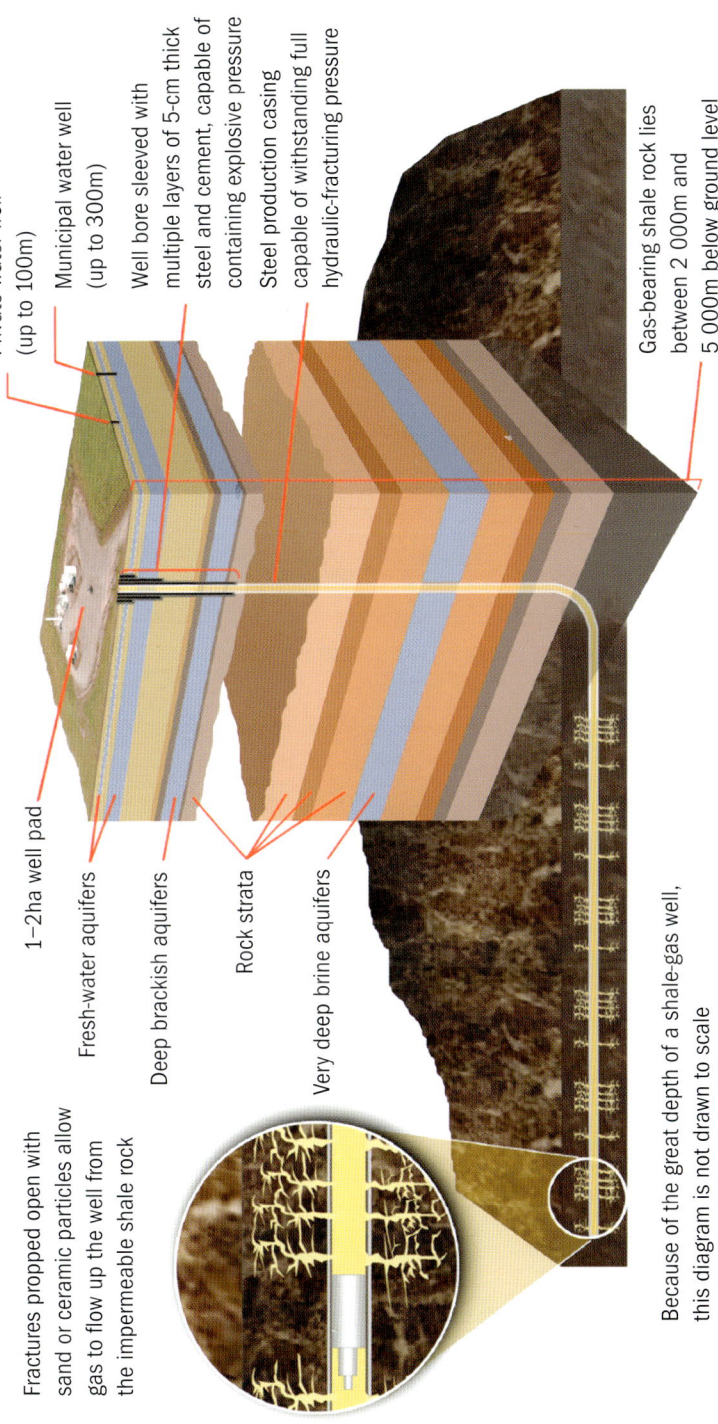

Private water well
(up to 100m)

Municipal water well
(up to 300m)

Well bore sleeved with
multiple layers of 5-cm thick
steel and cement, capable of
containing explosive pressure

Steel production casing
capable of withstanding full
hydraulic-fracturing pressure

Gas-bearing shale rock lies
between 2 000m and
5 000m below ground level

1–2ha well pad

Fresh-water aquifers

Deep brackish aquifers

Rock strata

Very deep brine aquifers

Fractures propped open with
sand or ceramic particles allow
gas to flow up the well from
the impermeable shale rock

Because of the great depth of a shale-gas well,
this diagram is not drawn to scale

Figure 1: Diagram of a horizontally drilled, hydraulically fractured shale-gas well, showing major features of the well, the surrounding geology and the
various kinds of groundwater through which it is drilled

Source: Adapted from 'Shale Gas: Applying Technology to Solve America's Energy Challenges', National Energy Technology Laboratory, 2011

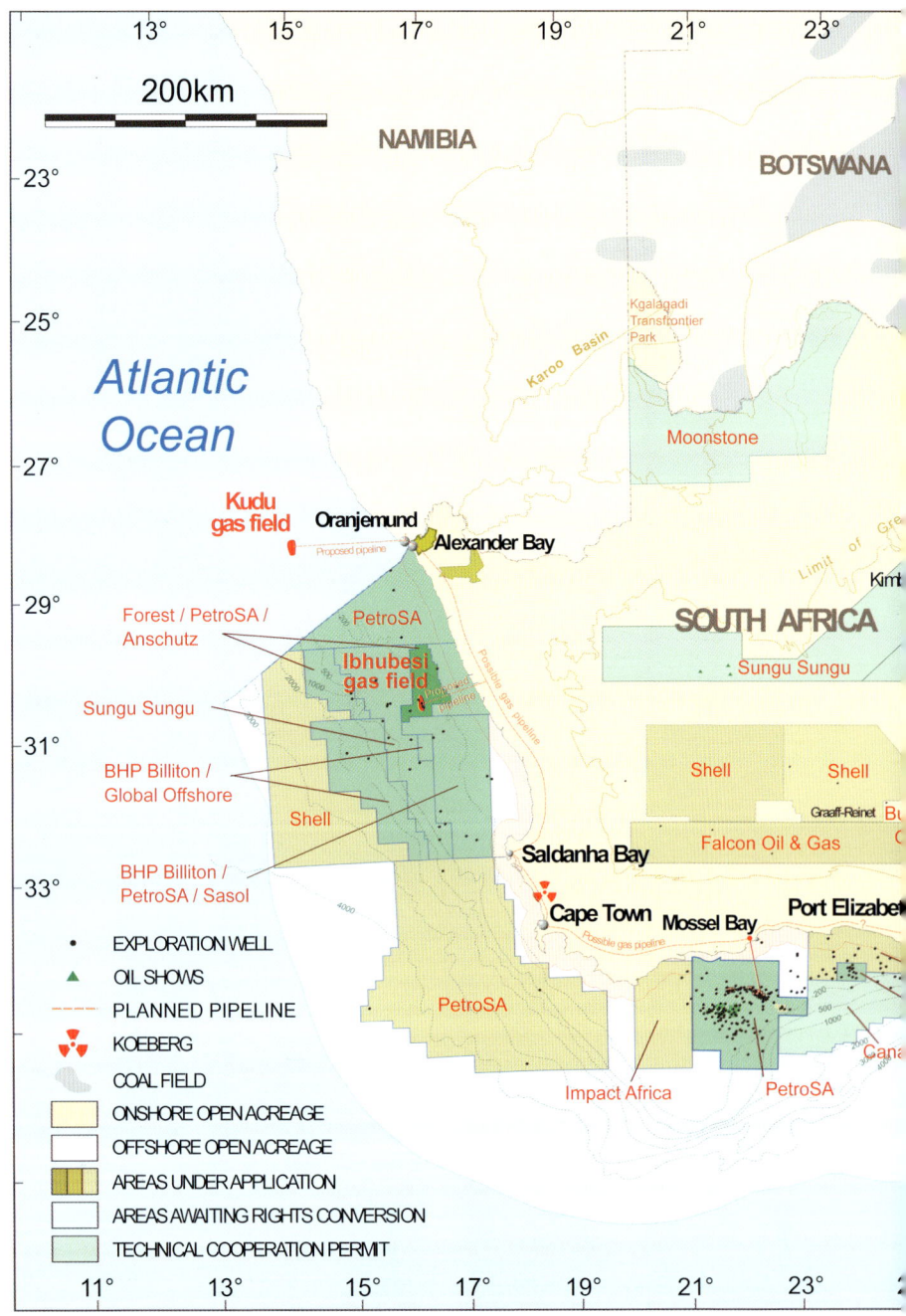

Figure 2: Gas- and oil-exploration projects in South Africa. Note the difference between technical co-operation permits (pre-exploration), exploration permits (pre-production) and full production rights

UCTION RIGHTS IN SOUTH AFRICA

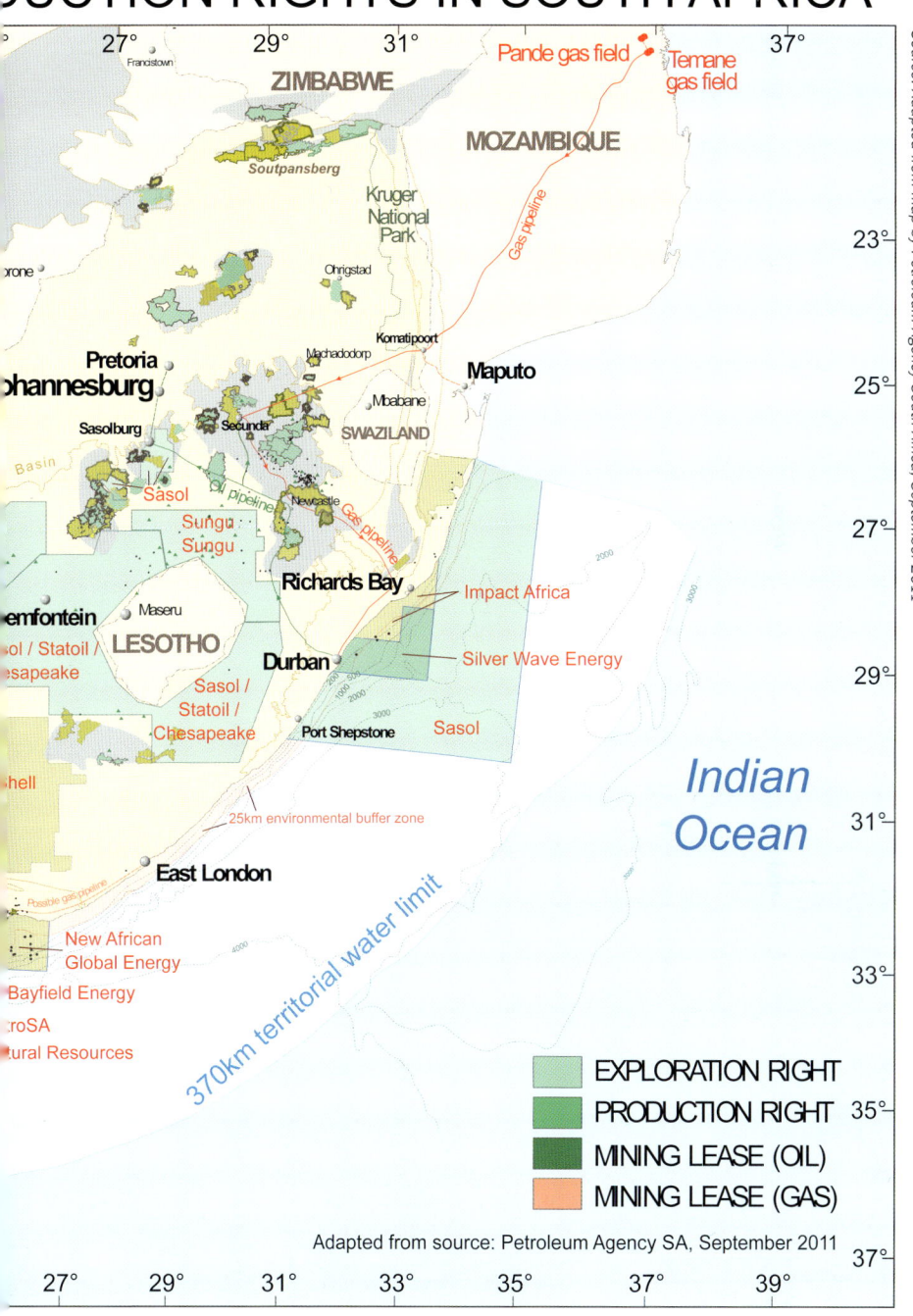

Source: Adapted from map by Petroleum Agency South Africa, September 2011

Francistown
27°   29°   31°   37°

ZIMBABWE

Pande gas field   Temane gas field

MOZAMBIQUE

Soutpansberg

Kruger National Park   23°

Ohrigstad

orone

Komatipoort

Machadodorp

Pretoria   Maputo   25°

ohannesburg

Sasolburg   Secunda   Mbabane

SWAZILAND

Basin   Sasol   Newcastle

Sungu Sungu   Gas pipeline

Pipeline

Richards Bay   Impact Africa   27°

emfontein   Maseru

LESOTHO   Silver Wave Energy

ol / Statoil /   Durban
esapeake

Sasol /   29°
Statoil /
Chesapeake   Port Shepstone   Sasol

hell

25km environmental buffer zone   Indian

East London   Ocean   31°

Possible gas pipeline

370km territorial water limit

New African   33°
Global Energy

Bayfield Energy

roSA

ural Resources

EXPLORATION RIGHT

PRODUCTION RIGHT   35°

MINING LEASE (OIL)

MINING LEASE (GAS)

Adapted from source: Petroleum Agency SA, September 2011   37°

27°   29°   31°   33°   35°   37°   39°

Source: *Checks and Balances Project*

HOW NATURAL GAS DRILLING CONTAMINATES DRINKING WATER SOURCES

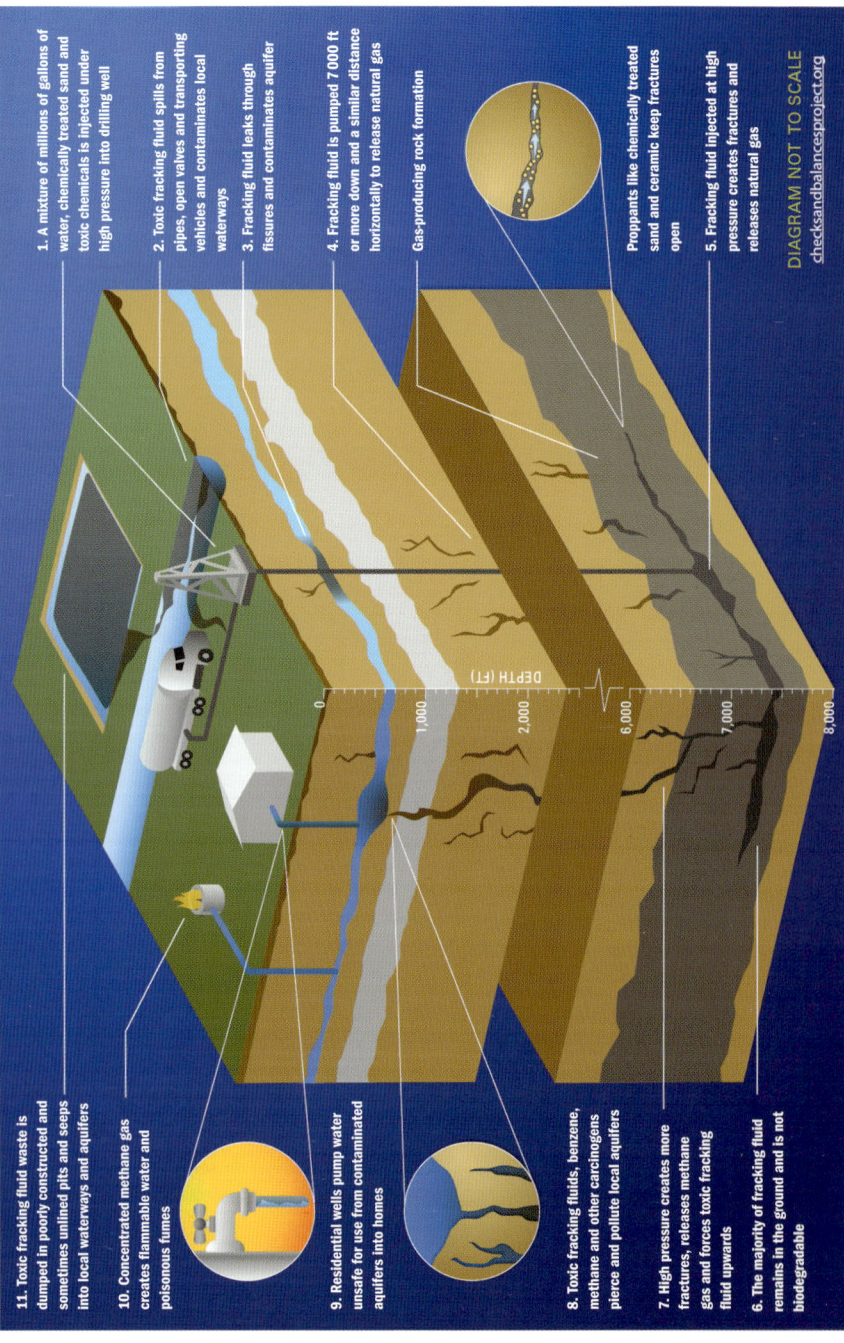

Figure 3: A very misleading diagram of what hydraulic fracturing is. Note in particular the absence of any well sleeving, and the spectacularly large fissures that supposedly reach through rock layers several kilometres thick

1. A mixture of millions of gallons of water, chemically treated sand and toxic chemicals is injected under high pressure into drilling well

2. Toxic fracking fluid spills from pipes, open valves and transporting vehicles and contaminates local waterways

3. Fracking fluid leaks through fissures and contaminates aquifer

4. Fracking fluid is pumped 7 000 ft or more down and a similar distance horizontally to release natural gas

Gas-producing rock formation

Proppants like chemically treated sand and ceramic keep fractures open

5. Fracking fluid injected at high pressure creates fractures and releases natural gas

DIAGRAM NOT TO SCALE
checksandbalancesproject.org

11. Toxic fracking fluid waste is dumped in poorly constructed and sometimes unlined pits and seeps into local waterways and aquifers

10. Concentrated methane gas creates flammable water and poisonous fumes

9. Residential wells pump water unsafe for use from contaminated aquifers into homes

8. Toxic fracking fluids, benzene, methane and other carcinogens pierce and pollute local aquifers

7. High pressure creates more fractures, releases methane gas and forces toxic fracking fluid upwards

6. The majority of fracking fluid remains in the ground and is not biodegradable

DEPTH (FT)

0
1,000
2,000
6,000
7,000
8,000

Figure 4: The Jonah natural gas field in Green River, Wyoming, incorrectly said to be representative of a fracking landscape

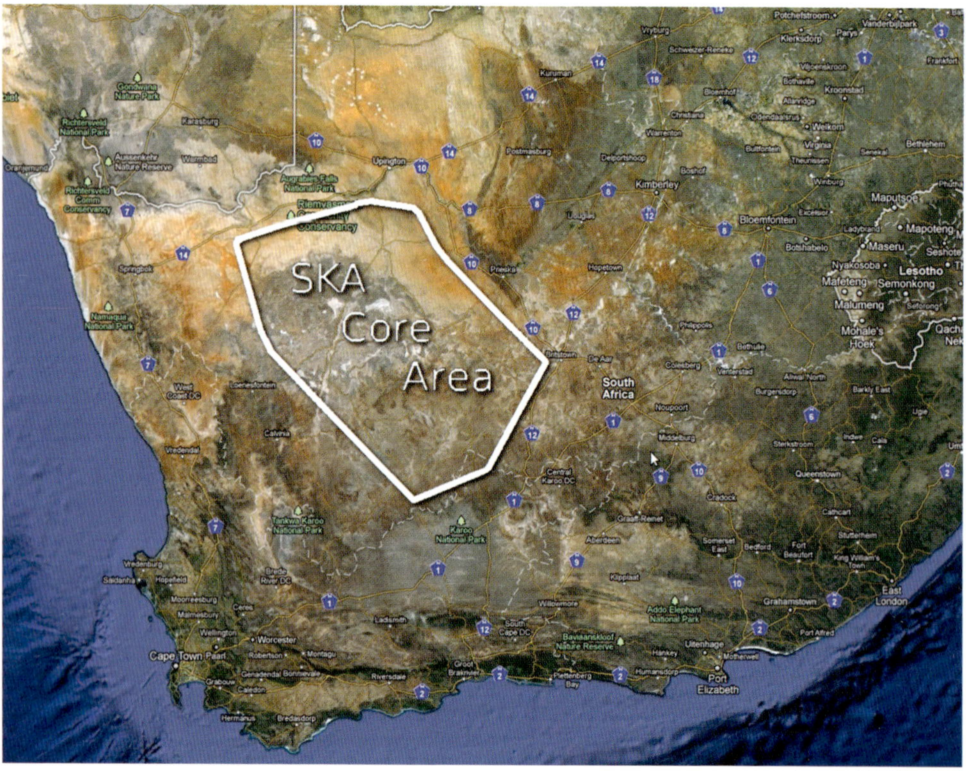

Source: SKA Africa and Google Maps

Figure 5: The Astronomy Reserve. A vast region in the central Karoo is protected by law from light and radio interference for the Square Kilometre Array (SKA) radio telescope project

Figures 6 to 9: Photographs of the KAT7 precursor array during and after construction, as well as a map indicating the scale of the core area of the Square Kilometre Array

Source: SKA Africa

Source: Adapted from images by SkyTruth and SKA Africa

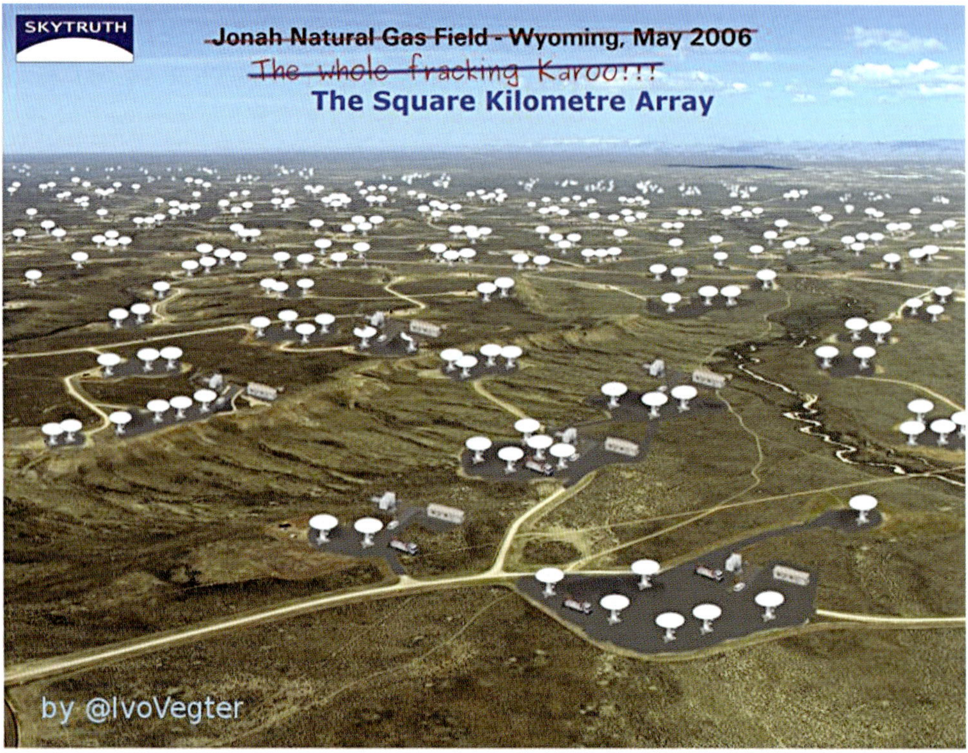

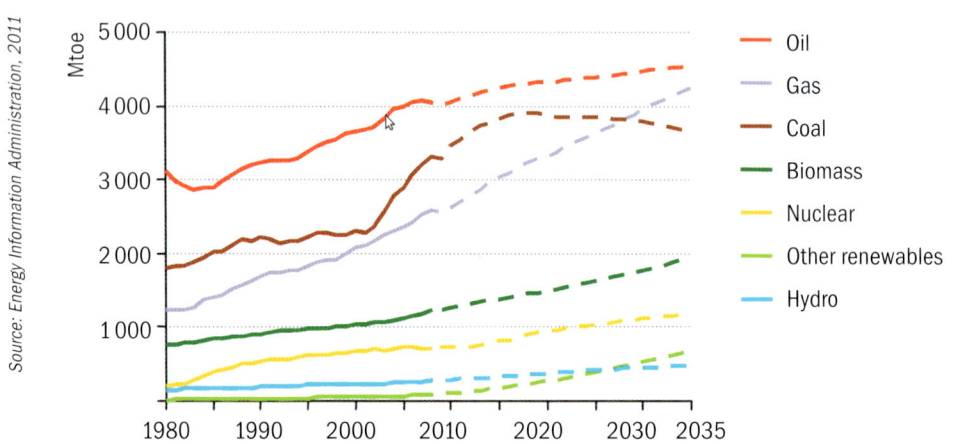

Figure 10: The author's impression of the SKA set in an arid landscape

Source: Energy Information Administration, 2011

Figure 11: World energy demand by type of fuel, past and future

An analysis of the oil remaining on beaches after six months of clean-up operations, conducted by scientists working for the US Coast Guard and published in February 2011, found that the situation was improving rapidly. The most dangerous chemical residue of the oil slick had been weathered and biodegraded by tide and time, leaving as little as 2 per cent of the original pollution on beaches that had not been cleaned. Risk to groundwater aquifers was considered minimal. Health effects, including cancer risk, from both short- and long-term exposure were found to be below acceptable risk levels set by the US Environmental Protection Agency (EPA). The report concluded that the environment would likely experience a greater threat from further clean-up than from the oil that still remained on the beaches.

All this from what even US President Barack Obama described as 'the worst environmental disaster America has ever faced'? Clearly he doesn't remember the dust bowls of the 1930s, despite the fact that his policies draw heavily from the Depression-era programmes of Franklin D. Roosevelt.

In an article in the *Telegraph* to mark the one-year anniversary of the accident, Ivor van Heerden, a scientist well known in the region for his work on coast and wetland restoration, is quoted at length:

'The spill was a disaster, but it was not the catastrophe that many people were portraying,' said Van Heerden, a marine scientist who once headed the Louisiana coastal restoration programme for the state's fragile eco-system of wetlands.

It was his intervention last July that first challenged the assumed wisdom in America that the spill was an apocalyptic environmental catastrophe.

'A lot of people, and that includes politicians and journalists,

did not want to hear the message that it was really not that bad,'
he said.

'But I went public as fishermen were committing suicide
because they were being told that this was end of their way of
life, that things would never recover. It simply was not the end
of the world.'[15]

The irony that environmental exaggeration was actually causing
human suffering and death is disturbing.

Writing for the *Houston Chronicle*, Bill King notes that alarming
videos of supposed 'underwater plumes' of oil, shown on television
news reports at the time and apparently shot by Philippe Cousteau
Jr, grandson of the famed explorer Jacques Cousteau, were not
likely to reflect their severity. Actual studies by the National Ocean
and Atmospheric Administration found that such plumes only
occurred at far greater depths than a scuba-diver such as Cousteau
could go, and at extremely low concentrations – a drop of oil in
a swimming pool. King adds that, by August 2010, the spill had
damaged only 1.5 km^2 of wetland, but that Louisiana loses wetlands
through both man-made and natural causes at a rate of 78 km^2
every year.

Almost exactly two years after the accident, Christopher Reddy,
the director of the Coastal Ocean Institute at the prestigious Woods
Hole Oceanographic Institution, offered a surprising and detailed
admission titled 'How science failed during the Gulf oil disaster'.
In it, he explained how scientists mismanaged their response to
the crisis and their contact with the media. The story of plumes
of hydrocarbons deep in the ocean, a matter of great scientific
interest, ended up being portrayed in sensationalist fashion as if
they were rivers of oil killing everything on the ocean floor. He
wrote:

> Academic scientists chose the research that most interested us, rather than what may have been most important to responding to the immediate disaster. We failed to grasp the mechanics of the media … It seemed so simple to us, but it was only newsworthy if the plume, at that time, could harm marine life or the environment.[16]

King and Reddy were not isolated contrarians. The respected (and frequently alarmist) *TIME* magazine also weighed in on the exaggeration, in an article by Michael Grunwald on 29 July 2010:

> The obnoxious anti-environmentalist Rush Limbaugh has been a rare voice arguing that the spill – he calls it 'the leak' – is anything less than an ecological calamity, scoffing at the avalanche of end-is-nigh eco-hype.
>
> Well, Limbaugh has a point. The Deepwater Horizon explosion was an awful tragedy for the 11 workers who died on the rig, and it's no leak; it's the biggest oil spill in U.S. history. It's also inflicting serious economic and psychological damage on coastal communities that depend on tourism, fishing and drilling. But so far – while it's important to acknowledge that the long-term potential danger is simply unknowable for an underwater event that took place just three months ago – it does not seem to be inflicting severe environmental damage. 'The impacts have been much, much less than everyone feared,' says geochemist Jacqueline Michel, a federal contractor who is coordinating shoreline assessments in Louisiana.[17]

When the mainstream media grudgingly concede a point to their arch-enemy on the right, Rush Limbaugh, the embarrassment can hardly get worse.

THE SEA IS FULL OF OIL

There are other reasons why the post-accident hysteria should have given a careful reader pause.

Deepwater Horizon was a major oil spill, and it raises serious questions both about spill prevention and about how best to go about pollution mitigation once a spill does happen. However, even this incident pales in comparison to some other oil spills.

Take wars, for example. During World War II, shipping losses were counted in lives lost and tonnage sunk, not oil-spill volume. In all the oceans in the world, a total of 7 800 ships were lost, including 860 oil tankers. Does anyone remember their parents or grandparents complaining about the catastrophic death of the oceans after the war?

Accurate numbers on what historians consider to be the largest ever loss of shipping from a single overall cause are not available. However, the Kuwaiti oil fires and other consequences of the 1991 Gulf War are better understood. Estimates of losses vary, but they range between 500 million and 2 billion barrels.

A great deal of this oil ended up in the Persian Gulf. It is uncertain exactly how much, but the UNEP estimates are at least double the amount spilled during the entire eighty-seven days of the Deepwater Horizon leak. Very little clean-up was undertaken, and along much of the Gulf's length, beaches, mangroves, lagoons and salt marshes were heavily oiled. About half the region's coral reefs and seagrass beds, which serve as breeding grounds for turtles and dugongs, were polluted. In addition, soot and oil droplets carried other pollutants, including heavy metals, into the water.

Needless to say, one would expect a catastrophic environmental impact from such a large spill in such an enclosed sea. That is, one might expect it if one reads the kind of coverage that Deepwater Horizon received. But a little rustle through the archives unearths a report from a research ship that surveyed the area for the UNEP a year after the war, looking for environmental damage. It found that much of the pollution had already dissipated. The 1993 report[18] noted much evidence of the pollution of 1991, especially in the northern region of the Persian Gulf. But taken together, the survey findings were not a devastating onslaught of heartbreak and despair.

The researchers concluded that 'very low concentrations of the oil were present in the samples from the water column', trace-metal content of the samples had not increased as a result of the spill, there was 'no evidence of large-scale sinking of oil' except in a few channel bottoms, and while sheltered bays remained oily, exposed beaches had recovered significantly thanks to weathering and erosion. The top layer of sediment was found to contain 'wide-spread, low-level contamination', but deeper samples contained only 'background levels' of oil. They said 'much of the residual oil' about one year after the spill 'showed considerable weathering'. The survey also was 'unable to reveal any demonstrable, direct effect of the 1991 oil release in most of the … area', and generally 'the coral reefs examined … appeared to be in good condition'. The limited reef damage that was found, the report noted, could also have been caused by other climate phenomena, such as the cold winter of 1991/92.

In the majority of the Gulf, south of Khafji on Saudi Arabia's eastern shore, 'other parts of the coral reef community, including fishes, macroalgae and invertebrates, showed no signs of unusual stress'. Spawning of the coral 'in the path of the oil' was also 'not impaired'. Several species of seagrass were tested, but there were 'no significant effects'. 'In general,' the report continues, 'the seagrass systems studied appear to be thriving.'

That even major oil spills don't leave a terrifying amount of destruction in their wake is overshadowed by an even more important fact: such accidents don't happen very often. In fact, ocean drilling is a trivial cause of ocean pollution. (See Table 5.1, which ranks the major sources of oceanic oil.)

Table 5.1
Sources of oil in the ocean

Natural seeps	46.2%
Consumption activities, including operational discharges	36.9%
Transportation, including spills	11.5%
Extraction activities	2.9%
Other	2.5%

Source: National Research Council, 2003.

Much more oil is spilled from tankers than escapes from oil platforms, but even tankers and rigs combined can't crack the 10 per cent mark, while all extraction and transportation sources combined account for less than 15 per cent of oil in the oceans. Moreover, as with production spills, the quantities spilled in the last decade or two are small and declining sharply. The best year of the 1970s was more than twice as bad as the worst year of the last decade. The trend in oil-tanker spills is a ringing declaration of the success of prevention efforts and safety standards, contrary to the impression one might get from the dramatic media imagery that goes along with any modern-day accident.

The vast majority of human-caused oil pollution in the oceans comes from operational discharges from vessels both in harbour and at sea, as well as fuel and oil disposed of by consumers that end up in sewerage systems. These spills are invisible and don't attract the cameras of environmental activists, but they are far, far worse than the Deepwater Horizon incident.

But here's the clincher. What no environmentalist will tell you is that the single biggest source of oil in the oceans is Mother Earth herself. Natural seeps from the ocean floor account for almost half of all the oil in the oceans. Such seeps have always been harbingers for oil prospectors, attracting them to oil-rich regions ripe for exploration, and the underwater world is no different from the plains of Texas or the deserts of Arabia.

In 1989, the tanker *Exxon Valdez* ran aground near Alaska, spilling hundreds of thousands of barrels of crude oil. This was nowhere near the largest ever oil spill, but it was a highlight for environmental activists because of the treasure trove of propaganda it produced in the form of photos of tar-covered beaches and oil-slicked birds. (As the late Constance Holden wrote in 1990 for *Science* magazine, spilled oil looks worse on TV.) Yet nature itself dumps an entire *Exxon Valdez* worth of cargo into the northern Gulf of Mexico every two years. Over the Gulf as a whole, 'the number is twice the *Exxon Valdez*'s spill per year, and that's a conservative estimate', satellite-sensing specialist Roger Mitchell of EarthSat told *Science Daily* in 2000.[19] Fortunately, the environment is a robust place, and the natural processes of bacterial biodegradation and weathering routinely take care of crude oil in the sea.

One of the most famous natural seeps lies off the coast of Santa Barbara in southern California. This spot is relevant, because it is here that the environmental opposition to oil drilling began.

DON'T SPOIL RICH PEOPLE'S BEACHES

Santa Barbara, California. The beaches of the rich and famous. This is where it all began.

This was the site for the first ever offshore oil-drilling project, at the turn of the twentieth century, and here, in 1969, began what has become the model for modern oil-spill activism. On 29 January, an oil platform known as Alpha, six miles off the coast, experienced a well blowout much like what happened to Deepwater Horizon.

The incident was then, and still is today, described as an 'environmental nightmare'.[20] This is the source of the archetypal photographs of oil-slicked birds treated at local zoos and of beaches cleaned up by armies of concerned residents, students, surfers, shopkeepers and children. Many of the techniques for dealing with the aftermath of an oil spill had their genesis on the beaches of Santa Barbara.

Today, nothing of the accident remains to be seen. The few lazy slicks that occasionally surface in the area come from the world's largest known natural oil seep at Coal Oil Point in the Santa Barbara Channel.

The environmental consequences of the Alpha blowout may have been exaggerated, but the consequences for environmentalism were huge. The Alpha platform disaster led directly to the first Earth Day, held later that same year. The first ever U2 'spy plane' photographs taken for peaceful purposes were of the Santa Barbara accident. It prompted a petition with 100 000 signatures to call for the banning of offshore oil drilling, and this was duly legislated for the next sixteen years. The Environmental Defense Center was founded, and the first environmental studies programme was established at the University of California at Santa Barbara. The US president at the time, Richard Nixon, signed legislation that created the EPA. The California Coastal Commission was established to regulate what people may and may not do near the beaches of the wealthy.

The backlash and activism of 1969 continues to reverberate around the world. Since 1969, the US alone has drilled over 50 000 offshore wells. Only fourteen significant spills have occurred, of which only three were major. Yet every time a Deepwater Horizon explodes, or a small area of a nature reserve is claimed for oil drilling, the outcry that was first heard in California sounds again, drowning out sensible policy and real-world concerns with exaggerated warnings of environmental doom.

All this is not to say that the Deepwater Horizon accident was not serious, or that BP, its contractors and the safety inspectors who lay down the rules did not fail on several levels. Oil companies and government regulators will learn from this event, and those affected have every right to demand clean-ups and claim reparations from those responsible.

The risks posed by industrial activity need careful monitoring as well as diligent mitigation. Productivity, prosperity and progress are fine things, but it is perfectly appropriate for environmental groups to warn that there's no need to be stupid, reckless or negligent in their pursuit. Keeping oil companies – and indeed everyone else – on their toes about safety and environmental risk-management is a noble aim.

THE RISKS OF EXAGGERATION

However, the consequences of exaggerating risks and accidents can be just as serious as the consequences of downplaying them. By overstating their case, environmental groups can do as much harm to people's lives, and even to the environment itself, as they accuse supposedly greedy oil company executives of causing.

The only difference is that the harm of such exaggeration is less easy to spot. It doesn't collect as a shiny slick on the surface of the sea, waiting for press photographers to show up. It doesn't lead to oil-coated seabirds that tug at the heartstrings of newspaper readers. But ironically, one example of the consequences of exaggerating risks is Deepwater Horizon itself.

Oil companies are perfectly capable of sustaining and even improving the protection of the local environment. A classic example is Rabi, an oil-drilling operation in Gabon, which is located right between two national parks, Loango and Moukalaba-Doudou. When a study of the region's biodiversity was conducted in partnership with the Smithsonian Institution, it not only documented thousands of hitherto unknown species and their habitat, but it found that the best-protected area was the forest directly around the oil wells. The rigs themselves, operated by Shell, occupy an area limited to a 100-metre radius, and the site was found to be even richer in biodiversity than the official national parks nearby. Oil workers are inducted into the ways of the forests upon arrival in the camp, ecologists and biologists are on the company's payroll to oversee scientific work, and site facilities are leased to local researchers for their own biodiversity projects.

True, an accident could happen that could damage the region, but the chances are small and manageable. Yet whenever there is a proposal to drill on sensitive land, especially in rich countries like the US, environmentalists exaggerate the dangers and downplay the benefits. They lobby energetically to shut down the plans. The same is true for proposals to drill near the coast, where local residents fear their beaches might be despoiled, and environmentalists have long backed them up with publicity campaigns, political lobbying and lawsuits. The result is that oil companies seeking new resources have to venture ever further offshore, or into ever more volatile countries, to be permitted to drill.

This is the very reason why expensive, super-sophisticated rigs such as Deepwater Horizon get built. However, operating a drill that only hits the ground 1500 metres below sea level, and then has to penetrate several kilometres of sand, sediment and rock before it reaches oil, is inherently more risky than tapping reserves in shallow waters or on land. The 'not in my back yard' phenomenon that started on the beaches of Santa Barbara, combined with the sensitivity about remote, unspoiled regions such as the Arctic

National Wildlife Refuge and the Karoo, perversely forces oil companies to accept ever greater risks.

A secondary consequence of the overzealous environmental movement is the high cost of energy extraction. Deep-water drilling is far more expensive than land-based or coastal drilling, and the costs to companies of complying with vast reams of rules and regulations, and running research projects and public outreach campaigns just to keep environmentalists happy, are enormous. Entire departments are employed for the sole purpose of overseeing environmental risk management and occupational health and safety.

These disciplines are, of course, necessary, and regulators are well advised to protect the rights, property and lives of citizens by requiring safety standards. However, if regulators are informed by exaggerated claims about risks by environmentalists, they are likely to over-regulate. Sometimes, as we have seen with deep-water drilling, the consequences are perversely worse than what the environmentalists were trying to prevent in the first place.

The added costs can only make matters worse for the world economy. It raises the cost of conventional energy – and ironically, that's exactly what alternative-energy boosters like environmentalists want, since their favoured sources are not cost-competitive. But this hurts the people who depend on energy for their food, their transport, their manufactured goods and their health care. The people of Santa Barbara might be able to afford solar panels and expensive wind power, but the people of Africa cannot.

Environmental groups such as Greenpeace have declared that 'a ban on all new oil drilling is the only way to avoid another spill disaster'.[21] This is true, of course. Just like a ban on all manufacturing is the only way to avoid more industrial pollution, and a ban on flying is the only way to avoid another air disaster. However, this is not a rational, wise and considered response. It does not signify a group that is serious about both human prosperity and the health of the environment. This is the position of a group that actively opposes human welfare, merely because it has the potential to cause environmental damage.

While it is proper for reporters and commentators to expose mistakes and wrongdoing whenever they happen, presuming evil intent on the part of BP because it is an easy target for popular prejudice is neither useful nor true. That throws the baby out with the bath water.

Going further, by demanding that oil drilling be halted altogether in the hope of preserving a pristine environment, is worse. It's like keeping the bath water and throwing out only the baby. Such a step will reduce living standards, impoverish the poor and, ultimately, cost lives.

Deepwater Horizon is the sort of accident that nobody likes to see, but to which resourceful people respond with alacrity, with a desire to minimise the damage and a resolve to prevent similar disasters in future.

The record shows that the oil majors are for the most part succeeding in reducing both the number and scale of accidents, be they tanker spillages or oil-well blowouts. They are also getting better at remediation, as the ecological condition of previous spill sites testifies. And they're learning to operate responsibly even in sensitive environments, as the story of the Rabi oil wells in Gabon demonstrates.

No doubt lessons will be learned from the Deepwater Horizon accident. The cynical might want to claim that changes will be made only for the sake of oil workers' lives, or even just to save oil companies the costs of lost equipment, clean-up costs, criminal penalties and civil suits. That's fine. That constitutes more than enough self-interest in doing what customers, regulators and the public at large demand.

THE EARTH IS FULL OF HOLES

If Mother Earth really were a living organism, as Naomi Klein claims, then perhaps another event that happened at the same time as Deepwater Horizon should have given us pause. While Klein was describing an underwater pinprick as a 'hole in the world', a rather larger hole was spewing out vast amounts of toxic fumes, green-house gases and particulate pollution.

That hole went by the delightful name of Eyjafjallajökull, a volcano that erupted for six days in April 2010, spewing vast plumes of glass-filled ash up to airline-cruising altitude and caus-ing the most serious air-traffic disruption over much of Europe since World War II.

If you made a cute Google mappy thing to properly scare chil-dren about the scale of the ash cloud, you'd have to compare it to your continent, not your city. This was a violent 'wound', gushing 'blood' into the atmosphere.

Granted, getting all self-righteous and over-the-top about natural disasters doesn't really fit in with the environmental narrative that humans are destroying the planet. From a public relations point of view, it wouldn't do to go around blaming the earth goddess for making volcanoes erupt. And it would have been just plain rude to make fun of Iceland for the unpronounceable name they gave to the volcano west of Mýrdalsjökull, just when its people were suffering harsh austerity measures because the country was busy defaulting on its debt. Priorities matter.

7

Fun with Facts and Fallacies

THE ART OF ARGUMENT

I argue – a lot. It's what I do. I argue with my friends. I argue with my mother. I argue at pub quizzes. With the quizmaster – even when our team is winning.

I've made many friends doing so, though I daresay I have lost a few who feared they'd end up throwing things at me.

I prefer to argue in writing, and I've been arguing online for years. At first, it was just a hobby to satisfy my interest in matters not directly related to the technology beat I used to cover as a journalist.

The most intriguing subject to me was economics. I needed to understand the late-1990s internet boom and its subsequent crash. I couldn't understand the popular waffle of the day about a 'new economy' and how the old rules about antiquated notions – like making money – didn't count any more. I was puzzled to find that few of the mainstream economics experts I was reading in the media were able to explain either the dot-com boom's rise or its collapse. Most supposed experts I'd read on the subject saw it coming, but only in hindsight.

This led me into an extensive period of study, during which I investigated all the major schools of economic thought, until I found one that most closely described the world I was watching. That this was not a formal education was, in a way, a benefit. I wasn't limited by the orthodox views that are routinely taught at universities, and was able to come at the competing schools with a fresh, critical eye.

My learning continued (and continues still) with history, since I had to understand not only the late-1990s boom, and the much bigger housing boom, but also the booms and busts before them. The Great Depression, the advent of central banking, the history of money, banking and financial regulation, the Reagan Revolution, the gold standard, the economic policy differences between countries and the impact of wars were all topics that fed into my ever-expanding body of knowledge.

So did the defining public concern of my lifetime, the environment. I was born less than a year after the first Earth Day, held on 22 April 1970, and grew up paging through a magnificent atlas published in 1972 that had large colour pictures – what we would call 'infographics' today – of major features of the earth. I vividly recollect the pictures explaining solar wind and the Van Allen belts, and the illustrations of environmental pollution and their impact on the water cycle and food chains. I recall the fears of the time: that the next ice age was upon us, and when it came it would freeze our farms with alarming rapidity. I recall that we risked precipitating this calamity by starting a nuclear war, and I always wondered why pictures of a frozen world under a dim nuclear-winter sky seemed to scare people more than having New York, London, Moscow and Amsterdam wiped off the map. Even then the notion of humanity as an undesirable plague on the face of the planet was in vogue, and even then it unnerved me for reasons I wouldn't have been able to explain. Still, a 'Campaign for Nuclear Disarmament' button seemed to express a rational political dream for a young teen in the early 1980s.

My discussions with friends and readers about technology and economics, progress and prosperity, would often segue into the very expansive field of environmentalism. The science and economics behind it soon became frequent items on my reading list.

I had enough background in mathematics, applied mathematics, philosophy of science and other university subjects to further these interests on my own, though I never did get my degree. If the fascinating subject of political philosophy had not distracted me from the more tedious parts of Politics 101, and if I could have turned off my brain long enough to learn the language of Marxism in Sociology 101, I might have been endowed with the proper papers, complete with Latin phrases, acclaiming my acceptance into the ivory towers of academia.

Instead, I took to earning a living by writing about technology, while learning and testing my knowledge not against examinations, but against colleagues, friends, subject-matter experts, Keynesian economists, radical environmentalists, corporate cronies, government apologists and assorted trolls in public debating forums.

Initially, I limited my arguments to electronic mailing lists and online bulletin boards. As my knowledge grew, my opinions developed and my command of the art of arguing matured, I started a blog. It proved to be moderately popular, not only in South Africa, but with readers from around the world. I'd often post arguments daily, and loved to engage readers in debate. I felt that if I was going to get on a soapbox and pronounce truths (or at least amusing falsehoods) to the world, I owed it to my readers to defend my positions. Besides, it was educational, challenging and fun.

It wasn't long before one of my editors proposed that I should take my arguments about these broader subjects into print, and do it for money.

It is easy to argue in print. You publish a column, and when someone writes a letter in response, you feel immensely thrilled that someone other than your editor and your mother read your stuff, and you feel influential for five minutes. Then, if the letter writer is wrong, you publish it with a withering retort:

> Dear Sir,
>
> RE: PANDA'S ARE CRYING AND DYING – STOP THE BAMBOO TRADE
>
> The plural of panda does not take an apostrophe before the *s*. Such callous disregard for even the most basic elements of style demonstrates conclusively that nothing I say will be able to convince you how thoroughly, ineffably wrong you are.
>
> Kind regards,
>
> Ivo Vegter
>
> *PS It's time to renew your subscription. We trust your pocket money will cover it. – Ed.*

In print, a writer can hide behind the postal system, which does not work because it is run by communists.

Online, however, you can't vet comments. You're exposed. One mistake, and you're toast. Tackle a big subject that evokes strong emotions, like hydraulic fracturing or rhino poaching or road safety, and your position is open to attack not only from readers, but from people who make a living being scientists, regulators or activists in that particular field.

The key to debating such controversial issues is not just having the facts on your side. Accurate facts and data are merely tickets to the show. The trick is to winnow out the many logical fallacies and rhetorical tricks used by advocates for one position or another, and then to try to avoid them yourself so you're not open to the same lines of criticism that you exploit in the opposing argument.

This might sound like a cynical view of debate about important subjects of genuine public interest, but it isn't really. It is how we try to approximate the truth, and determine the best ways forward in terms of public policy and private behaviour.

It is not unlike the scientific method, which requires that a hypothesis be falsifiable, and then encourages scientists to poke holes in it. As Stephen Hawking explained:

> Any physical theory is always provisional, in the sense that it is only a hypothesis: you can never prove it. No matter how many times the results of experiments agree with some theory, you can never be sure that the next time the results will not contradict the theory. On the other hand, you can disprove a theory by finding even a single observation that disagrees with the predictions of the theory. As philosopher of science Karl Popper has emphasized, a good theory is characterized by the fact that it makes a number of predictions that could in principle be disproved or falsified by observation. Each time new experiments are observed to agree with the predictions the theory survives, and our confidence in it is increased; but if ever a new observation is found to disagree, we have to abandon or modify the theory. At least that is what is supposed to happen, but you can always question the competence of the person who carried out the observation.[1]

This is not negativity or 'denialism' (a word about which I'll say more later). It is how theories are tested against empirical fact. The attacks do not weaken scientific theories, but strengthen them.

In the preceding chapters, you will have seen a number of these logical fallacies and rhetorical devices. Each time, they're designed to promote a favoured action, exaggerate a fear to which the public responds or conceal the vested interests of those who make the argument.

The Greeks once called the study of philosophy and rhetoric – the art of argument – *sophism* or *sophistry*. This perfectly honourable pursuit was soon undermined by skilled logicians and orators who were more inclined to use their skills to deceive than to arrive at *arete*, which translates roughly to 'virtue' or 'excellence'.

This development should not seem strange in the modern world. The perfectly honourable law profession, which exists to obtain justice for those who are wronged and to defend those who stand accused, employs a class of people that have much in common with ancient Greek sophists. Its practitioners are expected to know by heart great swathes of case law, to be expert at logical reasoning in order to interpret law in light of factual evidence, and to be capable of presenting the argument with finesse before a judge or a jury. Yet, the very fact that they have mastered facts, logic and rhetoric makes lawyers the object of scorn, jokes and distrust among many ordinary people.

This is not fair, of course, but it is hardly surprising when the only way to win a legal argument with a lawyer is to be well versed in legal philosophy, evidentiary fact and case-law precedent, and then to use the same standard of reasoning and rhetoric the lawyer brings to the table. Even more important is the ability to spot the abuse of statistics and the fallacies of logic that the lawyer employs. It doesn't take many embarrassments before an otherwise perfectly good case falls apart.

The same goes for political debate, economic argument and, indeed, environmental policy discussions. Activists and lobbyists on both sides often use fallacies and exaggerations to advance their pet causes. Being able to spot them is the most certain way both to evaluate how much of what they say is invalid and to argue against the rest.

WIPE THAT SMUG OFF YOUR FACE

I've collected some of the best examples of *ad hominem* attacks – those arguments that play the man, not the ball – in a little biography I like to use online:

> Columnist. Stirrer. A snivelling sycophant. A rotten little shill.
> A typical conservative liberal, with stupid white guy smug.
> Arbiter of idiocy.

'Columnist. Stirrer' is my own contribution. The rest is user-generated content. The last bit, 'Arbiter of idiocy', is there for ironic value. It was intended as a condescending insult by someone who happened to disagree with an opinion I expressed about some argument or policy, but I consider it to be entirely true, of course.

These phrases are all culled from comments on columns, or from responses in debates on social networks. Like most fallacious arguments, they attempt to achieve an effect in the mind of the audience. No reasonably intelligent person will expect an opponent in a debate to ponder an *ad hominem* allegation and then quietly concede the argument.

Simple name-calling is a trivial case. It is seldom successful in convincing anyone, and at best riles up those who have already taken positions on the matter. More often, it just makes the name-caller look childish.

Sophisticated name-calling, however, can be very effective. To note that someone is of a particular race, class, occupation, nationality, gender or religion, when this sometimes influences the opinions of members of such groups on an issue, can sway the reactions of

others. As long as one can offer a tenuous reason why such classification might be relevant, a powerful bias can be established in the
minds of an alarming number of people.

Here's a case in point: 'You're against culling elephants, or farming
rhino for their horns? Well, you would be. You're a conservationist.'
This is a perfect example of arguing the man. True, a conservationist may well be more inclined to abhor unnecessary killing of
animals, but that doesn't make him or her wrong. Conversely, many
conservationists believe that culling is part of a necessary environ-
mental-management programme, given the facts on the ground
in the modern world. Indeed, many conservationists believe that
hunting brings economic value to animals that exceeds that of mere
ecotourism, and which can fund the breeding or protection of the
animals they care about.

True, conservationists tend to be reluctant to admit what they
know will offend some tourists, fearing they'll get their eyes clawed
out over views their clients think are inhumane, callous or greedy.
But many a conservationist who has seen an overpopulation of
elephants destroy a game reserve, or seen how banning rhino-horn
trade has failed to do anything to curb the cruelty of poaching, will
gladly concede that trade, hunting or culling ought to happen.

The point is that the mere fact that a person is a conservationist
does not predict what his or her position about hunting might
be, nor whether this position is valid – meaning logically sound,
grounded in facts and likely to achieve the desired effect.

An argument about breast cancer is not more valid because it
is made by a woman who has it. It could equally well be less valid
because the woman in question has not studied the subject or is
too close to it emotionally, and uncritically accepts that long-term
exposure to food preservatives or low-level ambient radiation might
have caused it.

When Steve Jobs, the late CEO of Apple Computer, first discovered he had a rare and deadly form of cancer known as pancreatic neuroendocrine tumour, he turned to 'naturopathic' therapies for help. He reportedly tried a fruit-based diet, acupuncture and hydrotherapy, and even consulted psychics. When the disease progressed rapidly, he turned to conventional medicine, which is well supported by science and clinical trials. He underwent surgery, but the intervention came too late. He died in 2011. Was he more likely to have been right about alternative medicine because he had cancer himself, was highly intelligent, and had all the financial and research resources at his disposal anyone could desire?

Tim Minchin, an Australian musical comedian, wrote a line in what he describes as a 'beat poem' titled 'Storm', which goes: 'You know what they call alternative medicine that has been proven to work? Medicine.' Is Minchin wrong because he is a comedian and a singer? Or because he is Australian? Or because he is, horror of horrors, a ginger?

If I argue that smoking is harmful to your health and therefore you should not do so, but I also argue in favour of relaxing anti-smoking legislation, are my views right or wrong just because I happen to be a smoker myself? If a non-smoker made the exact same claims, would you dismiss the argument because that person can't possibly know enough about smoking?

How a person or their circumstances are described can profoundly influence how an argument is viewed. It establishes a cognitive bias in the audience. Yet in each case, the claim is about the person, not the argument, and is not relevant to the validity of what they have to say.

People who hold certain views or exhibit certain characteristics may also appear less sympathetic to observers. This is a well-known feature of court trials, in which witnesses are chosen and coached in large part on the basis of how sympathetic they might appear to jurors or magistrates.

Audiences are likely to believe someone who appears to be a caring, hard-working or responsible person. They are less likely to believe someone who appears to be a deadbeat or a hardbitten, profit-driven rationalist.

If a slumlord in a dirty tracksuit waving an eviction order says a tenant ought to pay the agreed rent, this statement is no less true because few people can bring themselves to like the fellow.

If the well-dressed and caring mother of a promising young athlete says the mystical hologram in a Power Balance bracelet (a faddish, expensive but ineffective piece of plastic sold as a perform-ance aid) will allow her to 'push [herself] beyond the 100% limit',[2] she is no less of a dupe. (This raises an interesting case, of course. People who wear Power Balance bracelets should not be surprised to find that others assume everything they ever said, thought or felt is wrong. That is an *ad hominem* argument, true, but let's just call it the exception that proves the rule. The bracelet maker has filed for Chapter 11 bankruptcy protection in the US, after losing a $57-million class-action lawsuit for false advertising. This came ten months after the company publicly conceded: 'We admit that there is no credible scientific evidence that supports our claims and therefore we engaged in misleading conduct.'[3])

Again, the point is that character or personal attributes in no way reflect on the truth or falsehood of what a person says.

'Stupid white guy smug'. Let's look at that phrase from my on-line bio. Since the person who wrote this comment doesn't know the result of the sole standardised intelligence test I ever sat, the first adjective is an opinion. It is true that I'm white, but noting this fact is *ad hominem* on the face of it. It may have some relevance to my socio-economic circumstances and cultural outlook in certain arguments, but it certainly cannot reflect on the validity or otherwise of my arguments beyond this. It is also true that I'm a guy, but the same observations hold as for my race. In most arguments, gender is an irrelevant observation. But 'smug'? Now there's a one.

It may well be true that I held a particular view – beats me if I can recall what it was all about – and was also smug about it. Or I might have held that view and only *seemed* to be smug. The thing is, whether I really was smug or only seemed to be smug, or neither, is only material to the argument if my debating position was also false. Looking smug doesn't make me wrong. If I am wrong, however, looking smug about it worsens my position and dramatically reduces my credibility. The illusion created by the accusation of smugness is that my position is also false. After all, there is considerable visceral appeal to the image of a righteous smackdown being delivered to a smug idiot. Attributing smugness to an opponent, therefore, can be a very effective rhetorical smear.

Being pedantic in defending yourself against allegations of smugness by refuting its relevance to an argument can also be entertaining and self-indulgent, and may bore your opponent to the point of surrender.

All these arguments hope to evoke in the reader what is known in psychology circles as the 'halo effect', where a particular physical or character trait biases others to believe or disbelieve the speaker. Marketers know this technique very well. Highly visible corporate social-responsibility projects have a distinct halo effect on how customers view the company. Conversely, the sense that a company errs in some way – be it by polluting the environment, dodging taxes, overcharging or underpaying staff – quickly reflects in customer perceptions of the company's product. This is a logical fallacy, of course. Why would a product be better or worse because a nice guy or a bad guy sold it to you?

YOU SNIVELLING SYCOPHANT!

A common *ad hominem* attack in environmental arguments is the charge of vested interests. The references to 'shill' and 'sycophant' are intended to imply this.

The term 'sycophant' has a very crude origin involving female anatomy. It has come to refer to a flattering toady, a grovelling lickspittle, a fawning lackey. Presumably, the word is meant to apply to someone who voluntarily defends a person, a government or a company whom the opponent, for whatever reason, dislikes. However, even if such an accusation were true, this would not invalidate the argument at hand. Only facts and reason can do that.

'Shill' probably comes from old circus argot, but can certainly be traced to the early twentieth century, when it was used to refer to a paid associate planted in a crowd to act as a decoy or foil for a gambler, auctioneer or snake-oil salesman. Modern use refers to a person who covertly works for, or gets paid by, a company or special-interest lobby to promote a sympathetic view.

In this sense, Andreas Späth, whom we met in an earlier chapter, could be described as a shill. He wrote a column disputing my views about the validity of the arguments used by some critics of hydraulic fracturing, while identifying himself only as follows: 'Andreas has a Ph.D. in geochemistry and manages Lobby Books, the independent book shop at Idasa's Cape Town Democracy Centre.'[4]

The first phrase – about Späth's academic credentials – is an appeal to authority, which we'll deal with later, while the second phrase – about his job at the book shop – seems harmless enough. It is only once you discover that he is also 'Earthlife's anti-fracking co-ordinator'[5] that you realise there's a reason to view his arguments in a somewhat new light.

But calling someone a shill without evidence to that effect is, frankly, defamatory.

Note, however, that between the two terms – 'shill' and 'sycophant' – there is no way at all to defend the object of your opponent's dislike. One term accuses you of being bribed to defend it, and the other is a fallback position that accuses you of being a soft-headed lackey who is too stupid to demand a bribe in the first place.

Rhetorically, the terms are very useful indeed. Logically, however, they have little incriminating value without evidence, such as that supplied in endnote number four.

LOOKIT'EM, POOR STRANDED POLAR BEARS

A very common way for lobbyists, advocates for a cause, and even the media to phrase a particular position is to make true statements about one or several events, and then derive stronger conclusions from them. We can refer back to an earlier chapter for an example: 'Some aeroplanes crash, and almost always a lot of people die. Therefore, a lot of people die in aeroplane crashes.' Or worse: 'Therefore, flying is unsafe.'

Neither of those conclusions follows from the premises given. However, because the premises themselves are perfectly true, it is possible to respond to the unjustified generalisation with a clever dodge: 'Are you saying aeroplanes *don't* crash?'

When industrial processes pose a risk to the environment, this kind of reasoning is used often. Opponents will point to specific cases where accidents happened, and generalise from there that accidents are either common or unavoidable or both.

One very well-presented anti-fracking website, www.dangersof-fracking.com, established by the makers of Josh Fox's film *Gasland*, contains several examples of logical fallacies. Take this statement: 'Contamination: During this process, methane gas and toxic chemicals leach out from the system and contaminate nearby groundwater.

Methane concentrations are 17× higher in drinking-water wells near fracturing sites than in normal wells.'[6]

There may be a few anecdotal examples in which such contamination occurred, but that does not justify the general statement made here, which implies that this always, unavoidably, happens. In fact, as we've seen, the truth is that such cases are rare, and many alleged cases of methane in well water are not caused by hydraulic fracturing at all.

The support that is provided in the second sentence is likewise insufficient. This claim refers to the Duke University study, also discussed in the chapters on fracking. Recall that the study could not conclude that the methane originated from fracking wells, that it is naturally present in many water wells, that it is not a health hazard in drinking water, and that it is especially likely to be present in water near where gas drilling would prove profitable. Recall too that the study was unable to find any traces of the chemicals associated with hydraulic fracturing.

Thus the claim is factually incorrect in all but one of its basic claims (that elevated methane levels were found in some wells close to drilling operations), and the conclusion, that fracking invariably causes contamination of groundwater, is an outright lie.

The fact that something happened once does not mean that it always happens, often happens or cannot be prevented. Pointing to data that prove something has happened proves nothing stronger than that it has happened. Anecdotal evidence is not data. It is not proof.

Think I'm being too harsh on this lot? Try this claim from the website for size: 'Contaminated well water is used for drinking water for nearby cities and towns.' This claim sounds absurd, and it is, but there are people who believe this stuff.

Journalists are fond of the phrase 'some people say'. It is a neat way around saying 'I think', which sounds biased and subjective. It's also a way to avoid having to specify just how many people do the saying. One? Two? Five? Fifty per cent of a large survey?

This kind of vague short cut is not always a problem. Sometimes the possibility of an event or the existence of an opinion is worth reporting in its own right. Sometimes 'many' or 'a few' are sufficiently specific for an argument to hold. However, a reader should always be suspicious that a writer means to imply more than what is factually supported.

This happens not only in written work. Images are very frequently used by environmentalists and the media as instances of this technique. A single picture of a starving African baby is meant to imply famine. Yet it does nothing to show the causes or extent of the problem. Is the case common or rare? Chances are it is the worst case (and therefore the best shot) the photographer could find.

Somewhere in the Ten Media Commandments it is stated that an oil spill has not happened until there's a proper photograph of a pair of dead, oil-covered birds. Then it is instantly a disaster.

Recall the bird counting that went on after Deepwater Horizon, described in Chapter 6. After extensive searching over many months, 8 000 dead or injured birds were found. But out of how many birds in total? Did no one stop to think that there may be 8 million birds in the Gulf, and losing 0.1 per cent of those, while too many, does not constitute a crisis of devastating proportion that deserves vast amounts of money and time to remedy?

A brilliant example of this kind of emotional, statistical and logical manipulation of the truth can be found in a nature photo that made the rounds a few years ago. In 2007, several newspapers published a picture of a pair of polar bears standing atop a small ice floe, dramatically expressing what looked like the effect of melting ice. 'They cling precariously to the top of what is left of the ice floe, their fragile grip the perfect symbol of the tragedy of global warming,' wrote the UK *Daily Mail*.[7]

The picture has been used in several environmental campaigns, including by the high priest of climate alarmism, Al Gore. It was used in several newspapers, including *The Times of London* and the 'newspaper of record', the *New York Times*. It was used on many advocacy blogs. A simple image search on Google turned up over twenty versions of the photograph online. 'Their habitat is melting ... beautiful animals, literally being forced off the planet,' Gore said with the photo on the screen behind him, according to a report in Canada's *National Post*. 'They're in trouble, got nowhere else to go.'[8]

On the face of it, the situation looks tragic indeed. But is the photo depicting a common occurrence? Does it truly represent the state of the melting ice caps and the peril of the bears?

In fact, there's a lot more wrong with the photo than simply exaggerating from anecdotal evidence. It was actively misrepresented.

One of the newspapers that used the photograph was the *New York Times*, in a story titled 'Science panel calls global warming "unequivocal"'.[9] A correction to that story notes that the photo was completely misattributed – location, photographer, date, the works. Nothing was true.

Correction: March 3, 2007
A front-page picture caption on Feb. 3 about polar bears floating on chunks of glacial ice, illustrating an article on a global warming report, carried incorrect information from the Canadian Ice Service about when and where the photograph was taken, and about who took it. The picture was taken in August 2006 in the Chukchi Sea, not in 2004 in the Bering Sea, farther south. The photographer was Amanda Byrd, not Dan Crosbie.[10]

Other than that, as the *Wall Street Journal* columnist James Taranto might say, the story was accurate. But even then, he'd be wrong. The Australian TV network ABC tracked down Amanda Byrd and asked her about the photo. It turns out that the photo was taken in the summer. Ice was *supposed* to be melting. Polar bears are known to be able to swim great distances, and these two were not far from the nearby shore. Byrd reports that they didn't appear to be in danger.

The confusion was complicated. Byrd was a marine biologist on a field trip. She took the photograph, simply considering it to be a striking shot, and gave a copy to Dan Crosbie, of the Canadian Ice Service, who was on the same trip. Several years later he passed it on to Environment Canada, a government agency, to illustrate an article, which in turn sent it to several news wires, which passed it on to newspapers. Nobody got paid, Crosbie got credited, and the location and date were lost in the mists of time. A spokesperson for Environment Canada told the *National Post*:

> It's just too cute to be true. You have to keep in mind that the bears are not in danger at all. It was, if you will, their playground for 15 minutes, you know what I mean? This is a perfect picture for climate change, in a way, because you have the impression they are in the middle of the ocean and they are going to die, with a Coke in their hands. But they were not that far from the coast, and it was possible for them to swim … They are still alive and having fun.[11]

Byrd considered legal action to enforce her copyright, and soon enough Gore opted to make the problem go away by offering to pay her for the photograph. She now charges for its use.

Byrd told the media that she believes in global warming, but does not believe her picture says anything about it. It served a very useful purpose to the environmental lobby, however. It was no more than a picturesque lie, and a perfect example of the kind of exaggeration that has great impact when published, but suffers little harm when a correction is printed a month later.

THE DR EVIL FALLACY

Whenever someone uses a big, scary number in an environmental argument, I can't help thinking of the Austin Powers film *International Man of Mystery* in which his nemesis, Dr Evil, the villainous mastermind who had recently awoken from years of cryogenic sleep, plots to hold the world ransom for ... one *million* dollars. A henchman points out to him the relative paucity of a million dollars in the modern world, upon which Dr Evil demands ... one *hundred billion* dollars.

This trick is so common it is truly tiresome.

The oil-slicked birds have already been mentioned. Eight thousand out of how many, exactly? Ten thousand? Then we have a crisis. Ten million? Not so much.

The *Gasland* website referred to earlier offers a neat variation of this trick, and combines it with another – using scary words: 'Up to 600 chemicals are used in fracking fluid, including known carcinogens and toxins such as ... lead, uranium, mercury, ethylene glycol, radium, methanol, hydrochloric acid, formaldehyde'.

Let's leave aside the fact that half of those supposed chemicals are actually products of the rock being drilled, not additives to the fracking fluid. They may still present an environmental or health hazard, of course, but already the claim is factually deficient.

Six hundred chemicals sounds like a lot, and it is. That's why drillers don't use 600 chemicals in fracking fluid. They use perhaps a dozen. It's true that they select them from a shelf of 600 *possible* additives. The exact composition of the fracking fluid depends on the nature of the particular well, but there is no well that ever contains 600 additives.

Then, consider the use of words like 'carcinogens' and 'toxins'. Toxic it may be, but ordinary middle-class environmentalists routinely pour hydrochloric acid into their ordinary middle-class swimming pools, together with chlorine, which is a chemical weapon of mass destruction used ever since World War I. Then they tell their vulnerable, defenceless little children to go swim in the resulting toxic cocktail.

Methanol is commonly found in hardware stores as 'wood alcohol', and is used as an additive in denatured alcohol. The latter is used to make French polish, which is a luxurious finish for pianos, dining-room tables and other furniture.

Ethylene glycol is common antifreeze, which we all put in our motor vehicles. It is also used in shoe polish, inks and dyes, some cleaning solvents and some wood preservatives. It's not the most pleasant stuff on earth, and is very toxic when ingested in large quantities. However, it breaks down in water, air or soil within days or weeks; low-level exposure (20 mg/litre) for a limited period (ten days) is considered harmless by US health regulators; and early diagnosis and treatment has been very successful in people who drink large amounts of it. When last did you see a hazmat squad in full protective gear attending to a car accident in which antifreeze leaked out and ended up in the storm-water drains?

Almost all chemicals are toxic in some dose. The context that is lacking when environmentalists warn of chemicals is how much of a chemical one has to be exposed to before it causes health concerns or environmental damage, and what the likely risk of such exposure is. You'll rarely find such detail in the propaganda issued by environmental organisations.

By the way, how do you think Josh Fox, the *Gasland* director, got to know about the 600 chemicals? Wasn't his claim that the evil energy companies refused to disclose them? To quote the propaganda site: 'Help support the FRAC Act (Fracturing Responsibility and Awareness of Chemicals Act) which would require the energy industry to disclose all chemicals used in fracturing fluid ...'

It would seem Fox's claim of non-disclosure by the industry is contradicted by his own propaganda. However, a rich mine of information his site is not. A rich source of logical fallacies, statistics abuse and rhetorical sleight of hand, yes, that *would* describe the productions of Josh Fox.

Here's another example of numbers that lack essential context: Greenpeace Africa has a plan to save South Africa. This one doesn't involve high-seas piracy, getting thrown in jail or dumping truckloads of dirty coal all over company driveways for our municipal workers to clean up. It is called 'The Advanced Energy [R]evolution: Greenpeace's energy blueprint for a sustainable future and green development'.

From an economic policy perspective there's a great deal to quibble with in the Greenpeace plan, but that's not the purpose of this book. Let's just take one number. Greenpeace says: 'South Africa can create around 150 000 new jobs in the energy sector in the next 20 years, and at the same time safeguard against catastrophic climate change – according to Greenpeace's new "Advanced Energy [R]evolution" report.'[12] So many new jobs – that sounds quite nice, doesn't it?

The South African government does the same thing. In its 'New Growth Path' policy document, one finds the following optimistic-sounding paragraph:

Technological innovation opens the opportunity for substantial employment creation. The New Growth Path targets 300,000 additional direct jobs by 2020 to green the economy, with 80,000 in manufacturing and the rest in construction, operations and maintenance of new environmentally friendly infrastructure. The potential for job creation rises to well over 400,000 by 2030.[13]

But there is no context at all for these jobs. In the simplest terms, anyone can create a given number of jobs if you subsidise them enough and you give them enough loan guarantees. Give me the money to pay them, and I'll set unemployed workers to work immediately, making stuff, cleaning stuff, or simply digging holes and filling them up again. In isolation, this is hardly a useful measure of economic benefit. You might as well give the workers cash and tell them to stay at home, if what they're going to do is not productive.

To determine whether a given job is a productive job, one needs more context. First, you need to know whether job losses in other sectors will reduce the apparent bonanza promised by Greenpeace or the South African government. Second, you need to know whether the projections are at all realistic. Third, if we're still expecting a net-positive jobs situation, we need to work out how much more consumers will end up paying for energy, because fossil fuels are being taxed and 'green' energy is being subsidised, or, worse, legislation makes it impossible to buy cheaper energy and consumers are limited to the kinds of green energy that is causing even subsidised companies to go bankrupt.

This is important, because the additional cost to the economy will, by necessity, reduce capital available for alternative uses, which would also create jobs – but productive ones. This qualification is important, because simply adding jobs is no different from simply raising any other cost. It benefits nobody except the supplier, at the cost of everyone else, including other suppliers.

Not having found the answers, I went looking in the full Greenpeace report, which promises 'a sustainable energy outlook for South Africa'.[14] Beautifully but inexplicably illustrated with images of indigenous Russian people and solar power stations in Spain, we find that the projection of 150 000 jobs is based on the second of two scenarios. In this scenario, the world would reduce greenhouse gas emissions by *80 per cent* by 2050.

Right. So 150 000 jobs won't happen.

What about the other scenario? The 'basic' one? Well, that still assumes a 50 per cent emissions reduction by 2050, which is only marginally less likely. But let's assume for the sake of argument that this scenario is achievable. What about jobs?

Oh dear. This is a problem. Right there, on page 61, in gloriously expensive technicolour, the real truth comes out. By 2015, the energy sector will employ 13 200 fewer people than it would have in the 'reference' (i.e. business-as-usual) scenario. By 2020, the lost jobs mount to 17 200. And by 2030, assuming that their projections are correct, they'll have created a whopping ... wait for it ... 300 jobs!

Doesn't that lovely big 150 000 number make you feel all warm and fuzzy now?

While you're feeling warm and fuzzy, how does the news that Ford saved itself $1.2 million and reduced CO_2 emissions by about 20 000 tons just by turning off PCs at night make you feel?

Pretty neat, not so? After all, 20 000 tons of CO_2 is 11 million cubic metres of gas. Visualise that, if you can. But let's add a little context, because what might look like delightfully big numbers are often more like Dr Evil's million dollars.

Ford's recent financial history has been like a roller coaster. In 2009, the company recorded a loss of $6 billion. In 2010, Ford made a profit of $868 million. In the first quarter of 2011 alone, the company's revenue was $33.1 billion. So, by turning off its PCs at night,

Ford saved 0.0009 per cent of one quarter's revenue. Its CO_2 saving amounted to 0.00007 per cent of carbon emissions from human activity, or less than a millionth.

Aren't we all proud of Ford now?

THE APPEAL OF APPEALING

Many forms of logical fallacy are an appeal to some property or quality that has no real bearing on the matter at hand, but predisposes someone to believe or disbelieve it.

A well-worn example is the appeal to authority. Because it is difficult to judge someone's expertise independently, or by verifying their reputation among peers, companies, governments and a host of other people resort to certifications, licences or academic credentials to determine expertise. This may be a useful shorthand, but letters behind a person's name doesn't make the person right.

Here are a few names. With letters, and then some:

Roger Morrison, M.D.

Bill Gray, M.D.

Dana Ullman, Master's in Public Health

Ronald W. Davey, M.D. (Physician to Her Majesty Queen Elizabeth II)

And here is what these people all believe in:

Homeopathy: (n) the belief that chemically pure water has 'memory', and remembers all the good stuff that's ever been in it, but not that it's been through a sewerage plant a dozen times. [With apologies to the aforementioned scientific expert Tim Minchin.]

Some experts genuinely do believe this, and others vigorously oppose this notion. One camp has to be correct, yet both camps speak from a position of supposed authority. A mere appeal to authority, on its own, cannot make a statement true or false.

In fact, since no one is ever always right, even the most respected experts – Albert Einstein, Isaac Newton, Adam Smith, Al Gore, Bono – must at least on occasion have been wrong, despite their authority.

A recent study of the retraction of scientific studies in a particular field found, startlingly, that although most scientists feel research misconduct is uncommon, 2 per cent of scientists report having committed serious research misconduct (such as falsifying results or fabricating experiments that were not conducted) at least once, and fully one-third admit to having engaged in questionable research practices. Authority does not trump mendacity.

Most errors in science are, of course, perfectly honest, and a perfectly natural part of the scientific process. For example, the world-renowned theoretical physicist Stephen Hawking used to think he was on a pretty solid wicket declaring that nothing could ever emerge from a black hole, an area of space-time with gravity so strong that not even light can escape. After all, the laws of thermodynamics weren't some woolly theory with spotty evidence. They were all-singing, all-dancing laws of nature. It turns out that Hawking was wrong. He had the good grace to revel in the discovery, and the scientific community had the good grace to call the stuff that comes out of black holes 'Hawking radiation'.

This anecdote touches on another fallacy – consensus. This is a popular one, especially when applied to the theory of man-made global warming. The term is usually used to mean that an overwhelming majority of qualified experts in a field believe a certain hypothesis, theory or prediction to be true, so therefore it is inarguable, especially by lesser mortals.

The first problem with consensus is related to authority: that of self-selection. Would one become a theologist if one did not already believe in a god? An atheist might dispute what a theologist has to say, but all qualified theologists would batter him with an appeal to consensus.

Likewise, would one become a climate scientist if one did not already believe in a climate crisis? Some might, but it seems likely that those who don't accept the idea of a climate crisis are more likely to stick to specialisations in which they feel they can make more meaningful contributions, such as geology, physics, statistics, computer science, hydrology, oceanography, applied mathematics or geochemistry, instead of joining the multidisciplinary world of climate research.

Therefore, that a majority of climate scientists turn out to agree that their subject is very, very important, while many of those who disagree are not climate scientists as such, should not be surprising. The mere appeal to consensus among climate scientists does not mean much, by itself.

There are other reasons to be sceptical when someone appeals to consensus. Does the claimed consensus really exist, or are the dissenters merely in the minority? In the case of global warming, there clearly are dissenters, so consensus is a bold claim, at best.

Does the consensus apply to what is being argued? Regarding global warming, the claim to consensus often applies only to the notion that average global temperatures have shown a rising trend for some number of decades. There are few people who dispute this, so appealing to consensus is to appeal to a trivial truism. There are numerous people, however, who dispute the extent of the warming, dispute the cause of the warming, dispute the likelihood of future warming predicted by computer models, or dispute whether or not it is advisable, or even possible, to do something about it. When someone claims that urgent government action at significant cost is not justified (as I do, for example, and will elaborate on elsewhere), an appeal to the consensus that temperatures have been rising is entirely beside the point.

Most importantly, does an appeal to consensus constitute a logically valid argument? That, it certainly does not.

There once was consensus that the sun orbited the earth. Sceptics were declared heretics, and were burnt at the stake.

There once was consensus, including among scientists, that light and radio waves had to travel through a medium. That medium was called aether, and it was believed to permeate the universe. But we now know that medium does not exist.

There once was consensus that disease was caused by something called humours, of which the human body contained four: black bile, blood, phlegm and yellow bile. These were associated with the four seasons, and with the four elements of earth, air, water and fire. This was accepted medical science for a period of 2 400 years – from the days of Hippocrates through the discovery of bacteria in the seventeenth century, and deep into the nineteenth century. It was only in the last 100 years that more than two millennia worth of scientific consensus was overturned for an altogether different consensus.

Of course, that does not mean that a consensus is always wrong, or even often wrong. Very often there's a good basis for the claimed consensus. However, an appeal to consensus on its own is not a logical argument. It is a fallacy.

The appeal to motive was touched on earlier, as a form of *ad hominem* argument. It is common for environmentalists to claim that those who disagree with them are motivated purely by greed and profit. Such a charge is low if it is untrue, and even if true, it doesn't make the views of such persons automatically wrong. Conversely, environmentalists are often motivated by the need to obtain research grants; to gain the approval of their peers, supervisors or governments; or to solicit donations for their lobbying organisation. Those motives are at least as mercenary as the profit motive is in companies, and can be just as corrupting.

A religious person may have a religious motive for arguing that one ought not to kill or steal. That does not mean they're wrong. They may also have a religious motive for arguing that one ought not to exert oneself on a Friday, Saturday or Sunday; that one ought not to eat pork, or any type of meat at all; that one ought not to have sexual relations with certain classes of people, be they the set of people who share one's gender, or the set of people one is not married to. They may well be right in some or all of these things, but their motive doesn't make it so. Equally, that they're right about not stealing out of religious conviction does not mean they're right about eating shellfish, or whatever it is that offends their religious sensibilities.

Another common error is to confuse correlation with causation. The Latin term for this fallacy is *post hoc ergo propter hoc*, which means because *b* happened after *a*, then *b* happened because of *a*.

A classic case of mistaken causation appeared in the scientific journal *Nature* in March 2000. It reported 'a strong association between childhood myopia [nearsightedness] and night-time lighting before the age of two: there were five times more children with myopia among those who slept with room lights on than in those who slept in the dark, and an intermediate number among those sleeping with a dim night-light.'[15]

Sounds plausible, right? The media, of course, ran with it. It was a lovely sensationalist story to scare parents about yet another perfectly ordinary comfort of modern life, and a lifestyle editor on a tabloid ought to be fired for not running it. Turn off the lights! You're hurting the babies!

As it happens, the scientists got their correlation and causation a little confused. When two events are correlated – that is, they are observed to happen together – either might cause the other, they might be totally independent and correlated only by coincidence, or a third factor might be the cause of both. The latter turned out to be the case here. Myopia is a genetic condition, and the reason children who ended up with myopia slept with the night-light on was for the benefit of their half-blind parents.

Newspapers aren't always right, even if they're highly respected, as is the *New York Times*. This was the paper that in 1920 mocked Robert Goddard, the father of modern rocketry, for proposing that a rocket powered by explosive charges could make it to the moon. In an unsigned editorial, the rapier struck:

> It is when one considers the multiple-charge rocket as a traveler to the moon that one begins to doubt and looks again, to see if the dispatch announcing the professor's purposes and hopes says that he is working under the auspices of the Smithsonian Institution. It does say so, and therefore the impulse to do more than doubt the practicability of such a device for such a purpose must be – well, controlled. Still, to be filled with uneasy wonder and express it will be safe enough, for after the rocket quits our air and really starts on its longer journey, its flight would be neither accelerated nor maintained by the explosion of the charges it then might have left. To claim that it would be is to deny a fundamental law of dynamics, and only Dr. Einstein and his chosen dozen, so few and fit, are licensed to do that.[16]

Note the multiple appeals to authority here: first, incredulity that the Smithsonian would condone Goddard's unscientific balderdash; and second, the licence granted to Einstein for the same. But the 'newspaper of record' hasn't finished digging its hole:

> That Professor Goddard, with his 'chair' in Clark College and the countenancing of the Smithsonian Institution, does not know the relation of action to reaction, and of the need to have something better than a vacuum against which to react – to say that would be absurd. Of course he only seems to lack the knowledge ladled out daily in high schools.

The editorial goes on to castigate Jules Verne, the nineteenth-century pioneer of science fiction, for having made the same error of believing a rocket would work in a vacuum: 'That was one of Verne's few scientific slips, or else it was a deliberate step aside from scientific accuracy, pardonable enough of him in a romancer, but its like is not so easily explained when made by a savant who isn't writing a novel of adventure.'

Hindsight is, of course, a perfect science, but still, to be so condescendingly and spectacularly wrong takes the kind of hubris only a few newspapers can muster. The *New York Times* did publish a correction, of course:

> On Jan. 13, 1920, [the *New York Times*] dismissed the notion that a rocket could function in a vacuum and commented on the ideas of Robert H. Goddard, the rocket pioneer, as follows ... Further investigation and experimentation have confirmed the findings of Isaac Newton in the 17th century and it is now definitely established that a rocket can function in a vacuum as well as in an atmosphere. The *Times* regrets the error.[17]

The correction was published on 17 July 1969, some twenty-four years after Goddard's death, forty-nine years after the offending editorial and one day after Apollo 11 had departed for the moon carrying Neil Armstrong, Michael Collins and Buzz Aldrin on a mission Jules Verne had envisioned more than a century earlier.

GODWIN'S LAW

Finally, I promised a word about 'denialism'. This is a particularly nasty form of the guilt-by-association fallacy.

The error is not to associate the person who makes an argument with a different opinion that person holds, such as by saying, 'Oh, but you support fracking, so you can't possibly care about the environment, so your opinion about rhino poaching must be wrong.'

It is not to associate the person with an organisation that can be attacked in lieu of addressing the argument, such as by saying, 'Oh, but you write for that self-confessed capitalist magazine [or, you work for that trade union], so your opinion about the government's New Growth Path must be wrong.'

It isn't even to associate a particular opinion with an exaggerated crime or fear, such as calling advocates of legal abortion 'baby killers', or referring to genetically modified plants as 'Frankenfood'.

This one is even worse.

One of the popular websites about logical fallacies is devoted to a particular cause. It is called The Nizkor Project, after the Hebrew word for 'we will remember'.[18] It is dedicated to the millions of Holocaust victims who died a gruesome death at the hands of Adolf Hitler and the Nazis. Its purpose is to combat the hatreds and prejudice that lead some to deny that the Holocaust happened, or to deny that it was as bad as most historians believe it to have been.

When General Dwight D. Eisenhower witnessed the horrors of the concentration camps, he ordered his staff to take photographs. He arranged visits by newspaper reporters, politicians and ordinary German citizens. He was afraid that future generations might not believe the extent of the horror that they found, and that some might one day claim it never happened. 'I made the visit deliberately, in order to be in [a] position to give first-hand evidence of these things if ever, in the future, there develops a tendency to charge these allegations merely to "propaganda".'[19]

Eisenhower was right. That tendency did develop. These people are called 'Holocaust deniers'.

Let's leave aside whether such people have a right, based on freedom of thought or speech, to hold and express their opinions, reprehensible and factually wrong though they may seem.

Until recently, the term 'denier' had not been used in any other context. Today, it is used to describe those who suspect that claims about man-made global warming are exaggerated by environmentalists.

In some cases, the comparison is made quite overtly: 'There are grave risks in drawing analogies with any aspect of the Holocaust,' wrote Peter Christoff for *The Age*[20] in Australia, before promptly doing exactly that:

> One easily oversteps the mark, losing a valid point amid counter-accusations of hysterical overstatement, of engaging in distressing, offensive and exploitative mis-association. Even so – and because of its resonance with Holocaust denial – the term 'denier' can be used to describe those who trivially reject the existence and threat of global warming.

There is a venerable convention on internet message boards known as Godwin's law. It says that the longer an argument continues, the more likely it is that one party will compare the other to Hitler or the Nazis. A corollary of this law states that this ends the discussion, and the party who invoked Hitler loses the argument. Fallacious as this law is, it has the merit of reminding participants in a discussion to be polite and to assume at least a basic level of humanity in their opponents.

'I use that analogy with great hesitation,' continues Christoff, 'but given what's at stake – the future of humankind rests on quick and uniform international action – it illustrates the immorality and potential damage of climate change denial.'

It clearly has not occurred to this sanctimonious fellow that many climate-change sceptics who are opposed to 'quick and uniform international action' are opposed on moral grounds at least as strong as those Christoff claims for himself. Some might feel that granting wide discretionary power over great swathes of human action to governments is a violation of the moral right to individual liberty. Others might believe that impoverishing people by spending trillions in the vain hope of achieving a speculative future benefit is deeply immoral and damaging. For people like Christoff (and he is not alone) to condemn all sceptics as equivalent to murderous Nazis demonstrates an intellectual arrogance that is beyond belief.

The term 'denier' connotes something that is self-evidently wrong, inhuman and evil. Using the term is a gravely prejudicial rhetorical technique on the part of environmentalists. It is also a logical fallacy. Whether or not Holocaust deniers are right or wrong has nothing whatsoever to do with someone's opinion on climate change. Trying to sway public opinion by suggesting otherwise ought to summarily condemn the environmentalists who so glibly use such prejudicial rhetoric to shout down critics of their exaggerations.

8

The Last Resort
Climate Change

THREE KEY QUESTIONS

By nature, people tend to be very committed to their positions, once taken. That is no surprise. For many, the coherent framework within which they make sense of the world around them is at stake. That's not to say that no coherent framework exists if one rejects the exaggerations and logical fallacies of extreme environmentalists and their political allies in the media. However, the act of conceding a point of view, accepting that one may have been gullible, or casting about for a different world view that makes consistent sense can be emotionally difficult and intellectually challenging.

It is not for nothing that younger people, who have yet to form settled opinions on many matters, often sound shrill and overwrought to their elders. Conversely, it is not for nothing that older people become fixed in their ideas, and unwilling to reconsider long-settled views and opinions.

Yet most of us, at one time or another, had the wrong end of the stick, and strongly believed in causes that later proved indefensible. Who didn't once tack sharply between political positions, as first one and then another appeared to be more popular, sported smarter T-shirts, had more emotive appeal or (if we were intellectually discriminating) boasted more solid arguments?

If it's hard for ordinary people to shake what psychologists call 'confirmation bias', which predisposes us to place more value on evidence that supports our preconceived ideas, and dismiss evidence against them, how much harder is it for environmentalists, who have placed their entire careers on the line to advocate a particular position?

It feels great to be idealistic, and to believe that your dedication to science or environmental activism can save the world. It is rather less romantic to be stuck studying an arcane field that involves lots of complex physics and mathematics and doesn't attract a great deal of popular interest, newspaper headlines, television interviews or grant funding.

The mating habits of Arctic terns, for example, is not a particularly interesting subject, unless perhaps you're a sailor in need of a supply of eggs. (In case you ever need to know, a tern will lay another egg if one of its eggs goes missing, so you can harvest them for food without a guilty conscience.)

If you want to make terns interesting to a few more people, link them to climate change, by writing papers such as 'Dispersal and climate change: A case study of the Arctic tern'.[1] Suddenly, they're a so-called indicator species that makes headlines in popular science magazines such as *New Scientist*: 'Arctic tern crowned "king of commuters" '.[2] You see, as long-distance travellers that migrate between the Arctic in the north and Antarctica in the south, terns depend on global wind patterns and krill populations in their summer feeding grounds, which in turn depend on climate, which, so the argument goes, depends on human activity.

Nigel Calder, a former editor of *New Scientist*, explained the phenomenon in the British documentary *The Great Global Warming Swindle*:

> If I wanted to do research on, say, the squirrels of Sussex, what I would do – and this is any time from 1990 onwards – I would write my grant application saying, 'I want to investigate the nut-gathering behaviour of squirrels, with special reference to the effects of global warming.' And that way, I get my money. If I forget to mention global warming, I might not get the money.[3]

Once the logical fallacies and exaggerations of some environmental argument or another have been pointed out, and the problem becomes uninteresting or insignificant, the fallback argument becomes simply: But what about global warming?

Indeed, what about it? By now, it should be clear that I'm not a fan of end-of-the-world alarmism, and the debate around man-made global warming – or 'climate change', as the more careful advocates of action would term it – is riddled with alarmism. A detailed discussion of the subject in all its complexity would take an entire book, and many good ones have been written on the subject. A few of them are included in the list of sources at the end of this book. Despite the widespread appeal to consensus, which is both factually incorrect (there are many dissenting or sceptical scientists) and logically unsound (consensus is not proof), the causes and effects of climate change are matters of ongoing debate among scientists. The bitter arguments fly to and fro daily in newspapers, academic journals and, indeed, entire books.

In the interest of cutting through the fever swamps, let's limit ourselves to establishing two important conclusions in relation to climate change. First, since this book is about environmental exaggeration, can we find significant cases of exaggeration – or better yet, admissions of exaggeration – among the scientists and bureaucrats who claim that governments must take urgent action to combat global warming? And, second, is there a relatively short, simple but logical answer to the climate-change conundrum? That is, can we take a position that doesn't stand or fall by detailed and comprehensive scientific argumentation about temperature-measurement networks, tree-ring proxies, ice-core records, the earth's radiation balance, solar activity cycles, sea-ice extent, urban heat islands, greenhouse-gas absorption rates, long-term variability in ocean currents, cyclical changes in the earth's orbit and axis orientation, the physics of cloud formation, and the reliability of computer modelling and forecasting?

An affirmative answer to the first question would suggest that there is as much reason to be sceptical of environmentalists' claims about climate change as there is to be sceptical every other time they cry wolf. A solution to the second problem would enable policy-makers and voters – most of whom are not scientists themselves – to rise above the convoluted complexity of climate-change controversy, and reach a sensible conclusion about how best to act.

As you may have surmised by now, this chapter sets out to show that the answers to both these questions are yes. So in the last section of this 'last resorts' chapter, we'll deal with a final significant question: Haven't environmentalists done a lot of good too? Shouldn't they be given at least some credit?

THE 'HOCKEY STICK' CONTROVERSY

In 1999, climate scientists Michael Mann, Raymond Bradley and Malcolm Hughes published a chart that set the world alight.

Accurate instrumental temperature records are very spotty around the world. Satellite observations of temperature, sea levels and sea-ice extent have only been possible since 1979. Even in those societies with the best scientific records, instrumental temperature data only go back 150 years or so. So-called proxy data, which are temperatures deduced indirectly from such evidence as tree rings or ice-core samples, have to be used to attempt temperature reconstructions prior to the age of modern instrumentation.

Such data, along with other historical evidence, show that the earth may have been quite warm for a spell in the early Middle Ages, during which period historians have noted a significant expansion of agriculture northwards in Europe. *Encyclopedia Britannica* describes the period as follows:

It has been estimated that between [the years] 1000 and 1340 the population of Europe increased from about 38.5 million people to about 73.5 million, with the greatest proportional increase occurring in northern Europe, which trebled its population. The rate of growth was not so rapid as to create a crisis of overpopulation; it was linked to increased agricultural production, which yielded a sufficient amount of food per capita, permitted the expansion of cultivated land, and enabled some of the population to become nonagricultural workers, thereby creating a new division of labour and greater economic and cultural diversity.[4]

Several centuries later, Europe suffered a particularly cold patch, centred on the early industrial revolution in the seventeenth and eighteenth centuries, when contemporary art showed that the river Thames routinely froze sufficiently hard for festive ice-skating scenes to take place in the city of London during winter. The *Britannica* defines what became known as the Little Ice Age this way:

[C]limate interval that occurred from the early 14th century through the mid-19th century, when mountain glaciers expanded at several locations, including the European Alps, New Zealand, Alaska, and the southern Andes, and mean annual temperatures across the Northern Hemisphere declined by 0.6 °C relative to the average temperature between 1000 and 2000 CE [Common Era]. The term Little Ice Age was introduced to the scientific literature by Dutch-born American geologist F.E. Matthes in 1939.[5]

Then along came Mann and his cohorts, and conveniently massaged the data to make these two phenomena – the first of which suggests that present warming is not unprecedented, and the second of which suggests that a 150-year temperature chart starts from a deceptive low – disappear. Like magic.

To the IPCC, the political group that had been formed at the United Nations to investigate the earth's warming and propose global regulatory countermeasures, Mann's chart was a godsend. Known as the 'hockey stick' for its distinctive shape with its sharp uptick in the twentieth century, the IPCC used it very prominently in its 2001 report, since it provided the best argument yet that recent temperature trends are highly unusual in a long-term historical context.

However, there was a problem. Two independent Canadian researchers, Steve McIntyre and Ross McKitrick, began to question the statistical validity of Mann's methods. Wrote McKitrick, who teaches economics at Guelph University: 'The particular "hockey stick" shape derived in the [Mann, Bradley and Hughes] proxy construction [of 1998] – a temperature index that decreases slightly between the early 15th century and early 20th century and then increases dramatically up to 1980 – is primarily an artefact of poor data handling, obsolete data and incorrect calculation of principal components.'[6]

Explains Richard Muller, a professor of physics at the University of California's Berkeley campus:

> [Michael Mann's method] tends to emphasize any data that do have the hockey stick shape, and to suppress all data that do not. To demonstrate this effect, McIntyre and McKitrick created some meaningless test data that had, on average, no trends. This method of generating random data is called 'Monte Carlo' analysis, after the famous casino, and it is widely used in statistical analysis to test procedures. When McIntyre and McKitrick fed these random data into the Mann procedure, out popped a hockey stick shape![7]

In the face of a spirited defence by Michael Mann and those who take his side in the man-made global-warming debate, several other scientists have since replicated McIntyre and McKitrick's results.

In his 2005 paper, McKitrick wrote:

> In the aftermath of the hockey stick's demolition, some scientists connected to the IPCC have tried to insist that it actually didn't matter that much to their case. Any such attempt to downplay the influence of the graph flies in the face of the print record. Without it the [UN IPCC's Third Assessment Report of 2001] would have been a very different document, it would not have been able to conclude what it did, nor could the IPCC have convinced world leaders to take the actions they subsequently took.

Around the world, governments included the hockey-stick graph in school curricula and promotional material aimed at winning public support for strong interventionist measures. Among the regulatory solutions was the international treaty known as the Kyoto Protocol, which implemented caps on greenhouse gas emissions and created tradable permits for use by major emitters of such gases in the hope of efficiently distributing the hoped-for reductions in emissions.

Although both the medieval warm period and the Little Ice Age have been restored to existence in the most recent IPCC report of 2007, they are mere shadows of their former selves, and the Mann hockey stick remains the image that the lay public most easily recognises as the graph that charts our doom.

This illustrates the grave impact that climate-change activism masquerading as science can have on government policymaking. The costs to an economy of measures such as the Kyoto Protocol, or any other taxes or regulations to limit the use of inexpensive energy in favour of cleaner but more costly alternatives, are immense. If policymakers base their conclusions on exaggerated data presented to them by a select group of elite scientists, while ignoring dissenting voices, the costs of their measures might not be justified. Because of these costs, many developing countries have lobbied hard to be exempted from measures to curb greenhouse gases. Frankly, even the rich world cannot afford them. How much more so is this true for countries that have yet to reach the level of industrialisation and prosperity that the West has achieved?

Like the agitation against DDT as a poison against mosquitoes, which began conveniently just as the last Western country (the Netherlands) was declared malaria free in 1971, the agitation against the supposedly harmful effects of carbon dioxide – the key by-product of industrialisation – just happens to have started once the West was already comfortably well off, but before the rest of the world has had a chance to catch up. For Western elites to now preach sanctimoniously against cheap forms of energy and industry in the poor world smacks of neocolonial discrimination, and the poor know it.

But are the activists right? In the years since Michael Mann made the halcyon days of the medieval warm period disappear, promising us not a new era of agricultural abundance, but a fiery future of desertification and weather disasters, the questions around the main scientific actors in the man-made global-warming narrative have only become more pointed.

A MASSIVE LEAK BLOWS CLIMATE SCIENCE WIDE OPEN

In 2009, a vast trove of email correspondence was released, either by means of an insider leak or via external hackers, from the East Anglia University Climatic Research Unit (CRU). This is one of the major academic centres for climate science, and it employs a number of high-profile scientists whose work supports the notion of man-made climate change and feeds into the political documents of the IPCC. An even bigger haul of internal documents was released two years later.

Sceptics called the incident 'ClimateGate', a term coined by *Telegraph* columnist James Delingpole. They argued that the leaked material proved unequivocally that climate research was a fraudulent conspiracy perpetrated by a close clique of scientists. This may have been an overstatement of what was found, but it wasn't entirely inaccurate. My own first reaction ran thus: '[V]indication is at hand for sticking to the conviction that alarmism about anthropogenic global warming was at least distorted, and probably an outright fraud.'[8] Soon, angered by the mendacity exposed by the leaked documents, I was to upgrade that claim to 'the outright fraud perpetrated upon a trusting world by a few dozen scientists at the University of East Anglia's Climatic Research Unit'.[9]

Several largely sympathetic investigations, most of which were conducted either by the universities that were implicated or the governments that funded them, cleared the scientists of outright scientific fraud. They limited themselves to very strict legal stand-ard of proof – namely, that participants in the email discussions deliberately lied.

One may easily concede that point and still make some alarming findings from a thorough reading of the leaked documents, as the in-vestigations for the most part did. These included that the scientists had deceived their governments, the broader scientific community and the general public, in part by illegally resisting lawful attempts by other scientists and members of the public to gain access to publicly funded research and data.

Among the most damaging revelations was that original source data for much of the official temperature record used by the IPCC simply do not exist any more. Scientists tried to explain away the exceedingly curious fact that they only retained adjusted data, claim-ing a lack of storage space prevented them from storing the original source data. However, this explanation is laughably flimsy in light of the trivial amount of space a simple temperature record would occupy on tape or disk even as long ago as the 1980s, as well as sug-gestions in the email trove that the source data had been available for use as recently as 2008.

All the leaked documents are available in searchable form. I have read many of the emails and some of the computer code in the archive myself. There is a great deal there to undermine any con-fidence one might have had that climate scientists are neutral researchers acting competently in good faith.

The emails confirmed McIntyre and McKitrick's critique of Mann's hockey-stick graph, and detailed how it had been created by grafting two different data sets to each other. One was derived from tree rings, which critics argue usually indicate rainfall, not temperature, and the other was a temperature record created with modern instruments. The weak underlying tree-ring data, which appeared to have been modified in a dubious way to generate a sharp uptick during the twentieth century, were exposed. So was the fact that where the tree-ring data and the instrument data overlapped, the two measurements contradicted each other. The solution? The proxy data were simply declared to be unreliable from 1960 onwards, and ignored. The process of explaining away the curious contradiction was described as 'Mike [Mann]'s *Nature* trick' to 'hide the decline',[10] which became the (slightly misleading) catchphrase of the sceptics who believed the world had been duped.

In a detailed examination of the tree-ring controversy, the author of *The Hockey Stick Illusion*, Andrew Montford, wrote that one of the scientists at CRU, Keith Briffa, found data in the tree-ring record that 'did not comply with the message that he wanted to convey',[11] and simply omitted it.

Another reason to disbelieve that the Mann hockey-stick chart is a very simple one: the problematic scaling of the vertical axis. If recent tree-ring data contradict instrument data, one cannot determine the proper scale for the proxy data derived from those tree rings. Selecting a scale or amplitude for the proxy half of the temperature reconstruction then becomes subject to arbitrary selection by the scientist creating the chart. This problem is illustrated in the evolution of the 'official' IPCC hockey-stick chart, which at various times shows the medieval maximum and industrial-era minimum at different scales or amplitudes, compared to the modern instrument record. Underestimating the amplitude of the older proxy data serves to exaggerate the more recent instrument data. Moreover, the impression of a very steep upward slope towards the end is likewise an artefact of the scaling on the vertical axis. In reality, the chart shows only fractions of a degree, so even if the chart were technically accurate and the proxy data are properly scaled, it is drawn up in such a way that it visually appears to show something much worse than what it really does show.

A lengthy piece of documentation, apparently written by a programmer named Harry, who appears to have had the unenviable task of cleaning up the data sets between 2006 and 2009, is a litany of complaints about what a mess they are. There is no version control, no consistency in file-naming conventions, no data integrity, no documentation; there is evidence that missing or inconvenient data are simply invented; and there is evidence that data are fitted to predetermined results rather than results being derived from the data. In the end, Harry simply admits that it is 'too late to fix it', so 'let's have a go at producing CRU TS 3.0! since failing to do that will be the definitive failure of the entire project'.[12]

Famed programmer Eric S. Raymond had a go at playing with the code that Michael Mann used to create the hockey-stick chart, and demonstrated that the hockey-stick shape was built into the adjustment code, so it did not need to come from the original (and now conveniently deleted) data. Comments in the code include 'Apply a VERY ARTIFICIAL correction for decline!!', before declaring a variable described in a comment as a 'fudge factor'. In Raymond's words, 'This isn't just a smoking gun, it's a siege cannon with the barrel still hot.'[13]

An email apparently sent by Rob Wilson, a research fellow at the University of Edinburgh, also challenged the validity of Mann's hockey sticks. Anthony Watts, a tireless documenter of scientific malfeasance in orthodox climate-change circles, quotes the email at length:

I first generated 1000 random time-series in Excel ... as these series are generally random white noise processes, I thought this would be a conservative test of any potential bias ... Using three different methods, I developed a [Northern Hemisphere] temperature reconstruction from these data ... The reconstructions clearly show a 'hockey-stick' trend. I guess this is precisely the phenomenon that Macintyre [*sic*] has been going on about.[14]

Another email, penned by Keith Briffa and posted on Steve McIntyre's blog, runs:

I am sick to death of Mann stating his reconstruction represents the tropical area just because it contains a few (poorly temperature representative) tropical series. He is just as capable of regressing these data against any other 'target' series, such as the increasing trend of self-opinionated verbiage he has produced over the last few years, and ... (better say no more).[15]

Yet another email, by Phil Jones, the director of the CRU, to Michael Mann, reads: 'Keith [Briffa] didn't mention in his *Science* piece but both of us think that you're on very dodgy ground with this long-term decline in temperatures on the 1000 year timescale [i.e., the stick portion of the hockey-stick shape].'[16]

It may be that the leaked code and emails do not conclusively prove fraud, if you suppose that the hard evidence – the original source data – is missing. However, they do appear to be more than just suspicious. They raise the question of why such a great deal of certainty was attached to a chart prominently used in political documents when there were extensive private doubts about exaggerations – both in limiting previous climate variability and in overestimating the rate of temperature rise in the twentieth century – that called into question the credibility of the conclusions.

As Muller wrote in his essay explaining the failure of the hockey-stick chart: 'Progress in science is sometimes made by great discoveries. But science also advances when we learn that something we believed to be true isn't. When solving a jigsaw puzzle, the solution can sometimes be stymied by the fact that a wrong piece has been wedged in a key place.'[17]

It is clear from many of the leaked documents that the level of certainty and unanimity among climate scientists is widely exaggerated, both by the media and by the public statements of scientists themselves. It reveals a world in which scientists are far from balancing the earth's energy budget, cannot account adequately for cloud formation, consider it a travesty that they cannot explain the apparent cooling of the years since 1998 and feel the need to shape their published conclusions so as to present a consistent message in the IPCC reports.

Some of the scientists fantasised about setting sympathetic journalists on McIntyre and McKitrick in an attempt to smear them, or even physically assaulting sceptics who disagreed with their conclusions. They also conspired to undermine the integrity of the peer-review system by lobbying heavily to keep academic papers that might provide ammunition to those who held sceptical positions on man-made global warming out of the science journals. In at least one case – that of James E. Saier, the editor of the *Geophysical Research Letters* (GRL) – it seemed the conspirators at the CRU may have succeeded in ousting the editor of a major journal.

Although the threat was made, and Saier indeed left the journal, he claims the two were not connected. Saier told Roger Pielke Jr, professor of environmental studies at the Center for Science and Technology Policy Research at the University of Colorado at Boulder, in an email published by Pielke:

> I haven't looked for, and don't intend to look for, my name in the CRU emails, but one of my colleagues did alert me to an email written by [Tom] Wigley in which he suggested that, if I were a climate skeptic, then steps should be taken to get me 'ousted'. Wigley's suggestion stems, I believe, from the publication of a *GRL* paper (by [Steve] McIntyre and [Ross] McKitrick) that criticized certain elements of Michael Mann's Hockey Stick paper. This paper caused a bit of a stir and because I oversaw the peer review of this paper, I assume that Wigley inferred (incorrectly) that I was a climate-change skeptic. I stepped down as *GRL* editor at the end of my three-year term, long after the excitement over the McIntyre and McKitrick paper had passed. My departure had nothing to do with attempts by Wigley or anyone else to have me sacked.[18]

That may be so, but that the matter was even raised as a reasonable and honest course of action casts a shadow over the integrity of the climate-science community.

There are many more instances of dubious scientific practices contained in the CRU leaks, but they're not the only source of evidence that global-warming claims are routinely exaggerated.

UNDERMINING THE 'CONSENSUS'

In early 2010, news reports emerged about what appeared to be a small error in the official IPCC reports about the fate of Himalayan glaciers. The region, often described as the 'third ice cap' for its abundance of high-altitude ice cover, might well be free of glaciers by 2035, the latest IPCC report of 2007 claimed, unless urgent action is taken to stop global warming.

As it turns out, however, the data in that report did not come from any of the thousands of scientific papers the IPCC claims to rely on.

It was a colourful claim concocted for a newspaper article, and has been roundly refuted by glaciologists, who say big glaciers take far longer to melt even in the most adverse circumstances. Rajenda Pachauri, the head of the IPCC, at first pooh-poohed the problem, but was soon forced to admit and correct the error. He told the British *Guardian* newspaper that the error had 'seriously damaged the IPCC's credibility',[19] but refused to apologise, saying that he had nothing to do with the error, and that in any case it was 'out of character' for the IPCC and doesn't change a thing about global warming.

However, Pachauri founded an outfit called The Energy and Resource Institute (TERI), which advises the government and sets national standards in India, and is itself involved in glacier research. It consults to companies who seek to comply with emission regulations or take advantage of carbon trading. It used the glaciers-will-vanish myth, complete with resultant widespread water shortages, in a funding proposal, claiming it was sourced from an authoritative study. Pachauri's organisation received £310 000 from the Carnegie Corporation of New York out of this proposal.

Pachauri was asked whether he knew about the error before the IPCC's meeting in Copenhagen in January 2010, and he told *The Times* of London: 'I became aware of this when it was reported in the media about ten days ago. Before that, it was really not made known. Nobody brought it to my attention. There were statements, but we never looked at this 2035 number.'[20] Funny, then, that *The Times* discovered that a journalist for the journal *Science* had asked him about the glacier claims in November the year before.

Pachauri has been inconsistent on whether or not he thinks there might be other errors in the report, at times flatly denying it, at other times conceding the possibility, but saying the chance is minimal. However, another prediction, which Pachauri has personally often used in speeches and interviews, is described by Christopher Booker in the *Telegraph*: 'One of the most widely quoted and most alarmist passages in the main 2007 report was a warning that, by 2020, global warming could reduce crop yields in some countries in Africa by 50 per cent.'[21]

The source of this alarming claim, writes Booker, is a report written for a Canadian environmentalist group by climate consultants who profit from global-warming fears. It, in turn, claims to have drawn on government reports from several North African countries, including Morocco and Algeria. However, the startling reality is that those governments actually expect the exact opposite. Algeria, for example, expects agricultural production to double by 2020. Only one report has found that in serious drought years cereal yields might drop to 50 per cent of the norm. In short, a worst-case prediction for worst-case conditions was cherry-picked for inclusion in the IPCC climate report, and given its official imprimatur as the way the world would look by 2020.

Many scientists have since distanced themselves from the IPCC and the scientists that it claims to represent in the political sphere. Among them is a prominent German socialist and environmentalist named Fritz Vahrenholt, who told *Der Spiegel* magazine, 'I feel duped on climate change.'[22]

The magazine cites his extensive credentials as a politician and environmentalist, and most recently as an executive in the renewable energy industry. Now, he aims to 'make enemies in all camps', claiming to want to break a taboo. 'The climate catastrophe is not occurring,' *Der Spiegel* claims Vahrenholt writes in his book *Die Kalte Sonne* (*The Cold Sun*).

He is not alone. Palaeoclimatologist Eduardo Zorita called for Michael Mann and others implicated by the CRU scandal to be barred from the IPCC process 'because the scientific assessments in which they may take part are not credible anymore'.[23] He adds: 'I may confirm what has been written in other places: research in some areas of climate science has been and is full of machination, conspiracies, and collusion, as any reader can interpret from the CRU files.'

Mike Hulme, working at the CRU itself, wrote that climate science has become 'sclerotic', and the IPCC 'may have run its course'.[24]

A frequently updated list of over a thousand peer-reviewed scientific papers that support sceptical arguments against climate alarmism is maintained on a website called *Popular Technology*, complete with detailed defences and rebuttals of criticism of the list. A US Senate Minority Report documents over 1 000 scientists who have officially signalled their dissent from what the IPCC continues to hold up as a consensus on climate change.

One of them, South African scientist William Alexander, professor emeritus of the Department of Civil and Biosystems Engineering at the University of Pretoria and a former member of the UN Scientific and Technical Committee on Natural Disasters, declared the IPCC and the climate-change issue 'dead' in a dramatic memorandum written in May 2010. He declares the cause of death to be:

1. Provably false assumption that human activities can influence global climate for which there is no scientifically believable evidence.
2. Provably false assumption that the increases in global temperatures are the cause of climatic changes. Multiyear variations in global climate are driven by variations in the receipt and poleward redistribution of solar energy via the atmospheric and oceanic processes, not temperature variations. This is high school physics.
3. Complete lack of numeracy skills and logical deductions by the climate change adherents.
4. Deliberate manipulation of climate change science to suit political objectives.[25]

These are strong words from a scientist – rash, even. But they are not surprising, given the record of exaggeration and underhanded backroom dealings among environmentalists and climate scientists.

Many years ago, the late Stephen Schneider, who had been professor of environmental biology and global change at Stanford University and a member of several environmentalist organisations, made two infamous statements that have haunted environmentalists ever since.

First, in 1971, he wrote a paper on the impact of atmospheric carbon dioxide and particulate pollution. In it, he notes that as carbon dioxide levels increase, their contribution to the greenhouse effect decreases, and that the 'dimming effect' of pollutants in the atmosphere is likely to cause a net decrease in surface temperatures. 'If sustained over a period of several years, such a temperature decrease over the whole globe is believed to be sufficient to trigger an ice age,' he wrote.[26]

His was one of several academic papers by respected scientists in the 1970s that predicted a coming ice age, and was also cited in early papers that began the U-turn in the middle of that decade towards predicting 'pronounced global warming'.[27]

Ironically, Schneider's paper gives the lie to modern claims that global cooling was merely newspaper sensationalism and was never seriously predicted by scientists. In any case, it seems Schneider was well qualified to make what was to become his most controversial statement, in 1989:

On the one hand, as scientists we are ethically bound to the scientific method, in effect promising to tell the truth, the whole truth, and nothing but – which means that we must include all the doubts, the caveats, the ifs, ands, and buts. On the other hand, we are not just scientists but human beings as well. And like most people we'd like to see the world a better place, which in this context translates into our working to reduce the risk of potentially disastrous climatic change. To do that we need to get some broadbased support, to capture the public's imagination. That, of course, entails getting loads of media coverage. So we have to offer up scary scenarios, make simplified, dramatic statements, and make little mention of any doubts we might have. This 'double ethical bind' we frequently find ourselves in cannot be solved by any formula. Each of us has to decide what the right balance is between being effective and being honest. I hope that means being both.[28]

Much water has flowed under the bridge since then, and perhaps Schneider was just a lone crank with an unresolved guilt complex over having been alarmist about both global cooling and global warming. Perhaps he really was misinterpreted, as he has always claimed. But there is a more recent case of a prominent environmentalist who came clean about climate-change alarmism.

James Lovelock is a fellow of the Royal Society, a highly successful independent scientist with close ties to the US National Aeronautics and Space Administration (NASA), and one of *TIME* magazine's thirteen 'Heroes of the Environment'.[29] He is perhaps most famous as the author of *The Gaia Hypothesis*, in which he posits the earth as a living organism of complex, interdependent systems. We'll have more to say about this notion in the next chapter, but suffice it to say for now that Lovelock has long been regarded as a sort of spiritual godfather of the environmental movement.

He has also been among the most strident believers in the theory of man-made global warming. To quote him in 2006: 'We are in a fool's climate, accidentally kept cool by smoke, and before this century is over billions of us will die and the few breeding pairs of people that survive will be in the Arctic where the climate remains tolerable.'[30]

And again, speaking to the *Daily Mail's* Sarah Sands in 2008: 'By 2040, the world population of more than six billion will have been culled by floods, drought and famine.'[31]

Writes Sands: 'Crackpot or visionary, the fact is that more and more people are paying attention to Lovelock, and that he, himself, supports the Intergovernmental Panel on Climate Change (IPCC) – the influential group who shared the Nobel Peace Prize with former American vice president Al Gore for their campaigns on global warming.'

'Humans are too stupid to prevent climate change,' Lovelock told the *Guardian* in 2010.[32]

It must be with considerable bemusement, then, that the followers of the nonagenarian environmentalist read the following on MSNBC.com in 2012: 'James Lovelock, the maverick scientist who became a guru to the environmental movement with his "Gaia" theory of the Earth as a single organism, has admitted to being "alarmist" about climate change and says other environmental commentators, such as Al Gore, were too.'[33]

Apparently, Lovelock still believes global warming is happening, but thinks he has been 'extrapolating too far'. He told the news outlet's Ian Johnston: 'The problem is we don't know what the climate is doing. We thought we knew 20 years ago. That led to some alarmist books – mine included – because it looked clear-cut, but it hasn't happened.' Referring to the fact that the last decade or more has seen an apparent plateau in global temperatures, he continued:

The climate is doing its usual tricks. There's nothing much really happening yet. We were supposed to be halfway toward a frying world now. The world has not warmed up very much since the millennium. Twelve years is a reasonable time … it [the temperature] has stayed almost constant, whereas it should have been rising.

According to Johnston, Lovelock isn't afraid to say he made a mistake, unlike a university or a government-employed scientist who might fear the loss of funding such an admission could cause.

A LOGICAL APPROACH TO CLIMATE CHANGE

In the face of exaggerated claims by scientists about truly complex scientific issues, what is a layperson to do? When we can neither ignore what might be a serious problem nor trust the technocrats and bureaucrats who turn out to be as fallible (and sometimes as mendacious) as any other human, how can we decide how best to live our lives or which public policy choices to support?

In a column written in March 2010, I revised an earlier article written in 2007, in which I proposed ten reasons to reject climate alarmism. The overall reasoning remains valid, and it has proved to be a useful way to think logically about the uncertainties.

The trouble with the theory of man-made global warming is that it depends on a number of links in a logical chain of reasoning. Cast doubt on even one or two of them, and the entire theory starts coming apart at the seams. Conversely, one can construct a logical argument that is robust, and that requires rebuttal on all of its points before it is in danger of collapsing. For the sake of brevity (and a better headline), the list was limited to ten items, but it can easily be expanded by another few points, if the need should arise.

So, aside from the fact that scepticism is a sound principle of scientific thought, why should one doubt the argument that governments have to step in with urgent measures to change human action in order to prevent, or at least mitigate, catastrophic climate change? Here's a formulation that has stood the test of time:

1. I'm not convinced that 'global warming' as a one-way bet is happening any more. *Au contraire,* I'm convinced 'climate change' is a trivial truism, warming and cooling periods are to be expected, and the most recent three-decade warming trend appears to have stopped [more than] a decade ago.

2. Even if it is happening, I do not believe that computer models are reliable predictors of future climate. *Au contraire,* the models are too incomplete, the input data is too scant, and both are too suspect to model a system as complex and chaotic as planetary climate with any confidence.

3. Even if they are reliable, I'm not convinced warming is a crisis. *Au contraire,* I suspect the environment has survived equally warm or warmer periods in the past, and that warming has brought mixed blessings, with benefits of which we could take advantage and drawbacks to which we had to adapt.

4. Even if it is a crisis, I'm not convinced human activity is the primary cause of climate change. *Au contraire,* I'm fairly sure our own contribution is small, and is dwarfed by the scale and natural variability of the climate system.

5. Even if it does, I'm not convinced CO_2 is the cause. *Au contraire,* although it is a greenhouse gas, paleoclimate records appear to show global warming causes higher CO_2 concentrations, not the other way around.

6. Even if it is, I'm not convinced the environment is so fragile that it cannot easily recover its equilibrium. *Au contraire,* I'm convinced the environment is a robust, stable system that can and does recover, even from significant damage.

7. Even if it is, I'm not convinced we're able to make significant changes to our carbon output. *Au contraire*, I'm convinced that even with the best of intentions and vigorous government enforcement we can make only small, woefully insufficient adjustments at the margins.

8. Even if we are, I'm not convinced that it will have a significant effect on the climate. *Au contraire*, I'm convinced even a drastic reduction in our CO_2 emissions won't halt whatever global warming may be occurring.

9. Even if it does, I'm not convinced we can afford it. *Au contraire*, I strongly suspect that significant cuts in carbon output will come at too high a price, especially for the world's poor, and we can better invest those scarce resources to solve more immediate problems.

10. Even if we can, I'm not convinced that telling people what they ought to do has any place in a free world, or that government's place is to enforce moral virtues by legislative force. *Au contraire*, I'm convinced that's fascism.[34]

The beauty of this line of reasoning is that one does not need to be an expert scientist or to be entirely certain of any of these points. You can let the eggheads – the people who enjoy debating the finer points of thermohaline circulation and cloud nucleation – argue them until the cows come home, but every single one of these statements would have to be falsified to sustain the claim that we urgently need coercive regulation, taxes and subsidies imposed by governments. If you were to flip a coin on each point – that is, attribute a 50 per cent chance that any of these ten statements is false – the chances that all of them are false, and consequently man-made global warming is a crisis requiring urgent action, would be one in a thousand.

That the entire edifice of the man-made global-warming theory was logically flimsy was enough reason to reject the popular position even before key elements of it were undermined by the leaked documents from the Climatic Research Unit of East Anglia University.

Thanks to the exaggeration, sloppy data handling and dubious ethics of some of the most well-respected climate scientists behind the UN IPCC reports, we now know that many of the core tenets of catastrophic global warming are in grave danger of collapse. If they do, they drag the entire edifice down with it. But even if they don't, it remains prudent to reject global-warming regulation in favour of vigorous economic development. Ultimately, an attitude of calm scepticism about climate change will have far greater benefits for the citizens of developing countries, as well as for a healthy and productive environment. After all, the world's poor depend on it to produce the wherewithal to live.

Meanwhile, back in the land of professional alarmists, there are, of course, many scientists who continue to maintain that man-made global warming is an urgent crisis that will cause catastrophe unless – or even if – we make urgent changes in the entire fabric of society. Among them is James Hansen, the global-warmist director of the NASA Goddard Institute for Space Science (from which dozens of former employees publicly distanced themselves not long ago). He recently penned an editorial for the *International Herald Tribune*: 'Game over for the climate'.[35]

Sure, Jim, sure.

GO ON, BE NICE

This leaves us with one final question. Do environmentalists serve no useful purpose at all? Is their penchant for exaggeration merely common, or is it universal?

It is tempting to dismiss everything environmentalists dream up and, frankly, they have only themselves to blame for crying wolf so often. It wouldn't be entirely fair, however.

Among their proudest achievements is the successful banning of chlorofluorocarbons – commonly known as CFCs, a kind of chemical commonly used in refrigeration equipment, air conditioners and aerosol cans. This substance, it was said, destroys ozone, a form of oxygen that forms an important shield in the earth's atmosphere against damaging ultraviolet rays from the sun. The solution was to phase out the use of CFCs and replace them with alternatives that would be less destructive to atmospheric ozone.

There is some controversy over the extent to which this rare claim of environmental success is true, but it would seem churlish to quibble. Let's just quote a *Washington Post* article, which reported in 2009 that the very solution to yesterday's crisis is contributing to the crisis of today:

> This is not the funny kind of irony: Scientists say the chemicals that helped solve the last global environmental crisis – the hole in the ozone layer – are making the current one worse … Now, scientists say, the world must find replacements for the replacements – or these super-emissions could cancel out other efforts to stop global warming.[36]

The *Washington Post* is wrong, of course. That actually *is* funny. For all the precautionary foresight to which environmentalists lay claim, their regulatory solution had exactly the sort of unintended consequences for which they berate the supposedly myopic, profit-driven capitalists who are responsible for society's economic development.

Another environmental success has been in the reduction of so-called acid rain. Although the threat was certainly exaggerated, with dark prophecies involving vast swathes of burned forests and sterile lakes, the reduction by international treaty and other forms of regulation in the industrial pollutants that cause acidrain did have a positive impact. A study conducted for the US Environmental Protection Agency in 2010 found that acid-rain programme benefits exceeded its costs by a factor of forty to one, which is undeniably a regulatory success.

Other areas in which environmental regulation has had limited success are in cases where property rights have not been, or cannot easily be, established. An example is the problem of depleted fish stocks. Because few countries have attempted a property-based approach to regulating fishing rights, fish suffer from what economists term the 'tragedy of the commons'. Because they are communally owned, nobody has an incentive to take responsibility for the sustainable management of fisheries. World Wide Fund for Nature programmes, like the South African Sustainable Seafood Initiative, hold out some promise that the pressure on threatened populations can be limited, although ultimate success in returning overfished stocks to sustainable levels has yet to be achieved. There is considerable empirical evidence from countries such as New Zealand, which has implemented a comprehensive system of tradable fishing quotas, that market-based solutions will ultimately prove to be more successful than environmental campaigns backed up by government-enforced regulations.

It is also fair to attribute to environmentalists a great deal of success in raising awareness of matters about which the public ought to know. But the same can be said for corporate advertising. We are routinely sceptical of advertising because marketing claims are often exaggerated, and we should approach environmental campaigns with the same degree of scepticism. More importantly, just as corporate influence can corrupt our politics, exaggerated environmental rhetoric becomes dangerous to our freedom and prosperity when we let it write our laws and regulations. We should no more trust environmentalists with public policymaking than we would trust loan sharks or oil barons.

Whatever the occasional successes, and however valid some environmentalist warnings might be, the frequent exaggeration of environmental campaigns has the dual consequence of undermining the credibility of environmentalists when they do call our attention to an issue of genuine concern, and of provoking regulatory overreaction on the part of governments.

In 1997, *The Economist* ran a story titled 'Plenty of gloom', which argued that 'forecasters of scarcity and doom are not only invariably wrong, they think that being wrong proves them right'.[37]

It cited many examples, some of which have been covered in this book. It noted only one scare in the previous thirty years that proved to have some merit, which was the effect of DDT on some birds and other predators. The story went on to say:

> Perhaps the reader thinks the tone of this article a little unforgiving. These predictions may have been spectacularly wrong, but they were well-meant. But in that case, those quoted would readily admit their error, which they do not. It was not impossible to be right at the time. There were people who in 1970 predicted abundant food, who in 1975 predicted cheap oil, who in 1980 predicted cheaper and more abundant minerals. Today those people – among them Norman Macrae of this newspaper, Julian Simon, Aaron Wildavsky – are ignored by the press and vilified by the environmental movement. For being right, they are called 'right-wing'. The truth can be a bitter medicine to swallow.

The magazine argues that exaggerating the world's problems is not better than underplaying them, because it results in ill-conceived government policies that either fail altogether, or impoverish people rather than making them better off. It also breeds a kind of fatalism and even misanthropy that should be distasteful to anyone who believes in the resourcefulness of free people and the sanctity of every human life.

In its conclusion, *The Economist* says:

> You can be in favour of the environment without being a pessimist. There ought to be room in the environmental movement for those who think that technology and economic freedom will make the world cleaner and will also take the pressure off endangered species. But at the moment such optimists are distinctly unwelcome among environmentalists … Environmentalists are quick to accuse their opponents in business of having vested interests. But their own incomes, their advancement, their fame and their very existence can depend on supporting the most alarming versions of every environmental scare.

I read that article in 1997 while sitting by a fire in a game reserve, enjoying the marvellous richness of Africa's natural heritage. It made a great impression on me then. Today, fifteen years later, it is a pleasure to recall it, and an honour to quote it as reflecting my opinion on the subject as well as I ever could.

9

The Church of Gaia

AN INQUIRY INTO MOTIVES

This chapter is polemical, and may offend some environmentalists. Of course, the same is true of the other chapters, which accuse them of routine exaggeration in the furtherance of objectives that are sometimes worthy, but are often questionable. But facts are sufficient to establish that many aspects of modern industry and agriculture, such as food production or energy supply, stir environmental movements and their allies in the media into a frenzy of scaremongering and legal obstructionism. The question that has to be asked is, why?

There can be no certain answers to this question. There can only be speculation, based on the nature of environmentalist arguments, and the kinds of things prominent environmentalists say.

There are, of course, many well-intended environmentalists who don't think too deeply about the practical consequences or logical implications of their beliefs. Many are merely interested in protecting what they believe to be a threatened environment, and many more raise the alarm about genuine issues of concern. However, the broader movement, and its core, goes well beyond merely caring about the environment.

Conversely, caring about the environment is not inconsistent with a belief in individual liberty and free-market capitalism, a belief in God or a belief that environmentalism, as a movement, is harmful to human progress and prosperity.

For example, in the view of environmentalists, it is not enough to extract oil or gas in a relatively responsible manner. Shell has to be kicked out of the country and fracking has to be banned, because the only alternative is the destruction of nature. Witness the Treasure the Karoo Action Group's page on Facebook: it does not ask for facts about risks and benefits, or aim to educate. It is titled 'chase SHELL OIL out of the Karoo! [*sic*]'.

Further, it is not enough to merely protect endangered species. They can only be protected in their original wild state, in their original location, and any habitat on which human development has encroached must be returned to a state of nature. The clock must be turned back; in fact, extinction is preferable to conservation in an 'unnatural' habitat (see the views of Priscilla Feral in the next chapter). Private game farming does not constitute conservation, one environmentalist recently told me. Hunting and trade in animal products cannot be accommodated in a sensible framework that benefits both people and conservation. No, they have to be stopped, at any cost.

It is not sufficient to eat responsibly and keep a watchful eye that your intake of things that could be unhealthy for you is not excessive. No, our bodies apparently are not capable of dealing with constituents of food that they don't need, so food has to be additive-free. Government must force corporations to do what the liver and kidneys did for our grandparents, and consumers must pay the costs of regulatory bureaucracy, even if they can't afford it and would, for example, be happy, for reasons of health, taste, poverty or self-sufficiency, to take their chances with the unpasteurised milk that most governments, including that of South Africa, won't allow any-one to sell.

It is not enough to minimise pollution. The industry that poses these risks must be shut down.

It is not enough to fret about the imagined consequences of genetic manipulation. Monsanto products have to be banned, even if farmers prefer them because of their genetic resistance to drought, bad soil conditions or pests, and their genetic predisposition to be more nutritious or to produce bigger yields.

Why do so many people find such extreme positions appealing? There are a few possible answers to this question. By nature, specu-lation about motives is just that: speculation. However, it is worth trying to penetrate the reasons why environmentalists are prone to exaggeration in order to discern their ultimate objectives, or at least the ultimate consequences of their beliefs. Sadly, the results of such speculation don't make their positions any more defensible.

THE RELIGIOUS LEFT

The first attempt at an answer comes from the late novelist Michael Crichton. He expressed an idea from anthropology: namely, that certain social structures reappear in all societies. One of them is religion. Humans have a psychological need to make sense of their own lives, and of their relationship to the world around them. If they reject God or some other formal religion, many still need something that gives their lives a sense of meaning or purpose, and that belief is often quasi-religious in nature.

As an atheist, I'm not sure I entirely agree with that premise. A true atheist is able to accept that there are limits to human knowledge, and that there is no reason to suppose that life has some purpose or meaning extraneous to itself. The atheist does not invent the notion of a supreme being, a pantheon of gods or some universal spirit to explain the things that are either inexplicable because they're not subject to reason, like love, or are yet to be explained by science, like the biological nature of consciousness, the evolution of DNA, or what dark matter is and where it might be found. In short, the atheist is willing to accept Shakespeare's dictum that there are more things in heaven and earth than is dreamed of in our philosophy, even as science advances and learning furthers our capabilities.

This does not mean that atheists are heartless, soulless beings, no different from animals, living only for survival and procreation. Their intellect will lead them to form sophisticated interrelationships with others, and these account for a host of apparently meaningful actions that go well beyond mere biological health or the pursuit of epicurean pleasure. Love and happiness are much more than just pleasure or survival, and one needs no deity to experience or pursue them.

Nor does this mean that because they have no external dogma against which to measure their behaviour, atheists are without morals. It is perfectly moral to act in furtherance of your own goals while granting others the freedom to do the same. It is a perfectly good standard of moral behaviour to believe that by furthering the interests of another, you may ultimately benefit the society within which you live, so that actions that might be considered charitable or social are not only justifiable but right. It makes perfect moral sense that a person cannot profit justly without doing some other person a good turn, whether by making them a useful product or providing them with a beneficial service. There's nothing supernatural about the moral injunctions against murder, theft or assault, or in moral codes that favour honesty, hard work, charity and humility. These qualities are necessary for a society to function effectively, and for people to cooperate in producing the wherewithal to live. From a merely practical perspective, such principles are moral because they promote life, liberty and prosperity. This practical reality explains why such laws resurface in all religious moral codes too.

However, when Crichton spoke generally of the psychological tendency to believe in some greater purpose beyond our understanding, it is hard to disagree. The wide variety of religions in all societies on earth provides strong circumstantial evidence for his position.

He went on to consider the relatively modern phenomenon of the secular urban elite. Crichton believed that far from being atheist, in that they reject as archaic and illiberal the strictures of traditional Western religions, many self-proclaimed secular humanists cling passionately to environmentalism as a religion. Indeed, the parallels he drew with the principles of Judaeo-Christian belief are startling.

Modern science, technology and commerce – eating from the tree of knowledge – have led humanity from a state of innocent purity to one of moral degeneration. If there once was an Eden, an unspoiled nature, it is now corrupted by our sins, our lack of care and our hedonistic excesses. As a consequence of this sin, we will be judged in an inevitable apocalypse of global warming or even human extinction. There is a hell, and we're in the handbasket that's headed there. However, we can assuage our deep guilt by seeking salvation in sustainability, practising good deeds, like recycling and partaking of the sacrament of organic food. We have a priestly caste, which includes Al Gore, James Hansen and Paul Ehrlich, who preach fire and brimstone, prophesy our doom and instruct us how to live our lives to attain salvation. We have saints, in the likes of Rachel Carson and Thomas Malthus, who came before us to prepare the way.

In Crichton's formulation, the deity is Gaia, the Earth Mother, and he calls environmentalism 'a perfect 21st century remapping of traditional Judeo-Christian beliefs and myths'.[1]

The Judaeo-Christian God is anthropomorphic – that is, the

deity shares features of living humans. Likewise, Crichton's notion of Gaia as Earth Goddess meshes with the idea made popular by James Lovelock: the earth – which Lovelock also calls by its ancient Greek name, Gaia – has the properties of a living organism. Lovelock explicitly invokes religion in his explanation of the Gaia hypothesis:

> Most of us sense that the Earth is more than a sphere of rock with a thin layer of air, ocean and life covering the surface. We feel that we belong here, as if this planet were indeed our home. Long ago the Greeks, thinking this way, gave to the Earth the name Gaia or, for short, Ge. In those days, science and theology were one and science, although less precise, had soul. As time passed this warm relationship faded and was replaced by the frigidity of the schoolmen. The life sciences, no longer concerned with life, fell to classifying dead things and even to vivisection. Ge was stolen from theology to become no more the root from which the disciplines of geography and geology were named. Now at last there are signs of a change. Science becomes holistic again and rediscovers soul, and theology, moved by ecumenical forces, begins to realise that Gaia is not to be subdivided for academic convenience and that Ge is much more than just a prefix.[2]

Crichton is aware that this conception of environmentalism as a religion means that there is no logical way of talking people out of their environmental beliefs. These are matters of faith, not science. If you've ever argued against religion with a Christian, you'll soon realise that there is no way past the assumption that because 'creation' and 'creatures' exist, a Creator must exist, or the circular logic that God exists because God's Word says so. No religion is subject to rational discourse divorced from blind faith.

Like Crichton, it is not my purpose to defend my religious views or lack thereof, nor to disabuse people of their own faith. They're welcome to it, provided that they do not impose it on other people. However, just as there are good reasons to divorce church from state, lest the people we elect to serve us in government assume God-given power over us or dictate that we act in ways our conscience does not permit, there is a good argument to be made that environmentalism ought to be divorced from government. By all means, advocate additive-free food, an end to hunting or a preference for renewable energy. However, when those ideas become imposed on others by the coercive power of the state, through bans of things people choose to do, or mandates for things people choose not to do, problems arise.

Besides imposing on human liberty, these ideas have consequences. It may be unjustifiable for you or me to kill an animal that is endangered, but when that animal is a poor person's only source of food or threatens his or her small farm, do we have a right to impose our prejudices on that person? And even if we do, is it reasonable to expect that our neocolonial orders will be obeyed?

When renewable energy is expensive, can we justify requiring the poor to pay their share of the cost of it, whether by forgoing tax expenditure on more urgent infrastructure, or by charging a high price for electricity or fuel?

When the West paid for its economic prosperity by suffering the pollution of industrial development, can we deny the same benefits, at a reduced but essentially similar cost, to the millions of people in the developing world that have yet to match the West's prosperity?

Crichton might not deny people their faith, but he does observe that many of the core myths of environmentalism are hard to sustain in the clear light of science. For example, the notion of the noble savage living in the idyll of a bucolic past is invoked ritually by the Green Movement, but evidence suggests that such an idealised past did not exist. It is a romanticisation, in the same way that the mental picture we have of sailing ships is far more elegant than the brutal reality of harsh labour, vicious punishment, grave dangers and physical deprivation that was required to sail them.

Pre-modern tribes and kingdoms engaged in frequent warfare of a particularly vicious kind, razed forests to build houses and ships, hunted species to extinction, enslaved workers and suffered much from disease. Some tribes were cannibals. Others imposed restrictive forms of 'taboo' that incurred penalties of death, or made human sacrifices to appease angry gods. Life expectancy was never more than forty, and was often as low as twenty-five. Child mortality was a matter of routine. Every family could expect to bear several children who didn't make it to the age of five, and several more who didn't make it to productive maturity. One out of six women died in child-birth. People died routinely of infectious diseases that today are rare, or rarely cause death, such as influenza, tuberculosis, bubonic plague and scarlet fever. Just to scrape together enough for a meagre survival required long days of back-breaking labour. Malnutrition was common and the little luxuries of life that we take for granted today either didn't exist or were reserved for a very small number of wealthy aristocrats.

Nature doesn't take care of its own. There is no idealised state of living in harmony with nature, and there never was. You adapt, or you die. You disrespect nature at your peril. Yet environmentalists routinely advocate a 'return to the land'. Instead of teaching commercial farming methods using modern technology, they preach subsistence farming and how to grow organic food using nothing more than wind and solar power. They seem to have this vision of the poor as an underclass that toils to produce what only the rich can afford, while living from hand to mouth like medieval serfs. It's an idle delusion on the part of the rich but incredibly harmful to the poor, whom they doom to dependence on the state and who remain vulnerable to the vagaries of a harsh and unforgiving nature.

Moreover, just as the notion of an environmental Eden has no basis in fact, the environmental prophets of impending doom have a dismal record. Some, such as *Population Bomb* author Paul Ehrlich, continue to predict that 'population growth itself now retards development, widening the rich-poor gap and increasing the distress to those left behind'.[3] This didn't happen. In fact, the first Millennium Development Goal of the United Nations, halving poverty by 2015, has already been reached. And it had nothing to do with foreign aid, but everything to do with the freedom to pursue economic development and global trade, which saw its most dramatic expansion in China, but has reduced poverty in all regions of the world.

In 1962 Rachel Carson predicted a 'silent spring', and in 1979 Norman Myers predicted 'the sinking ark'. In his eponymously titled book, he said that by the year 2000, some one million species might have gone extinct, at a rate of 40 000 species per year, amounting to 20 per cent of all the species on earth. It didn't happen. And there's a good reason for that: Myers's prediction appears to be entirely made up. He offers no factual evidence at all.

The Danish statistician Bjørn Lomborg quotes Myers's argument in his book *The Skeptical Environmentalist*. Myers began by quoting a 1974 conference, which estimated current species loss at about 100 per year. He continued:

> Yet even this figure seems low … Let us suppose that, as a consequence of this man-handling of the natural environments, the final one-quarter of this century witnesses the elimination of 1 million species – a far from unlikely prospect. This would work out, during the course of 25 years, at an average extinction rate of 40,000 species per year, or rather over 100 species per day.[4]

By the time former US vice president Al Gore got around to repeating these statistics as fact in his 1993 book *Earth in the Balance*, the extinction of 40 000 species per year was engraved in the popular mind. Yet it derives from no empirical sampling, no statistical extrapolation and no scientific observation. It is entirely rooted in a supposition. True, if we suppose the loss of a million species in twenty-five years, that would be 40 000 species a year. That's a trivial truism. If we suppose a shape to be a circle, it would be round. If we suppose we drink six beers, we'd consume over two litres and be mildly intoxicated. If we drive 100 km/h, we'd cover 600 kilometres in half a day.

If we suppose anything, restating that supposition in a different form does not amount to useful scientific truth. It is still only a supposition. Myers engaged in circular logic to move from an entirely arbitrary supposition to what is now taken to be common knowledge.

Actual data published by the International Union for the Conservation of Nature (IUCN) show that there is little evidence for mass extinctions as a consequence of habitat destruction. It appears animals are capable of migrating to more amenable areas when their original habitat comes under pressure. Even today, the IUCN says that 'actual extinctions remain low'.[5] Vertebrates are best documented, albeit still imperfectly, and when we extrapolate their known extinction rate to all other species, by assuming that they suffer the same rate of extinction, we still only get a rate of 0.2 per cent per quarter-century. This may be cause for concern, but it is a far cry from Myers's alarmist prediction in *The Sinking Ark*.

In the late 1960s, Paul Ehrlich ominously and famously predicted that 'the battle to feed mankind is over', and that millions would die of starvation in the 1970s and 1980s. It didn't happen.

Ehrlich had been an obscure lepidopterist, studying butterflies at Stanford University, but then he extrapolated the idealised mathematical models developed by Thomas Malthus from population growth in insects, and published a book. *The Population Bomb* exploded (sorry) upon the world in 1968. It made hysterical predictions of exponential population growth that would inevitably lead to catastrophic resource depletion and mass starvation. As if the coming ice age wasn't stress-inducing enough in that benighted decade, Ehrlich's book advocated radical population-control measures. The apocalyptic overpopulation vision was reinforced by a bestselling book published in 1972 by the Club of Rome, an international think tank, titled *The Limits to Growth*, in which the collapse of human development and population was predicted as a result of exponential growth in demand for limited resources. The failure of these prophecies of doom hasn't changed the authors' minds. They went on to write further books, renewing their end-of-the-world predictions, the most amusingly titled being *Beyond the Limits*, published in 1992.

Yet resources stubbornly refused to be exhausted. The date of 'peak oil', like the doomsday predictions of the crazier religious cults, keeps getting postponed.

In 1980, Paul Ehrlich accepted a bet offered by economist Julian Simon, who said that resource scarcity was not a crisis. The two agreed that market prices were an adequate proxy for scarcity, and Simon let Ehrlich pick any five metals, of which the pair bought a $1 000 basket to hold for ten years. If their prices rose, Simon would pay Ehrlich the gain. If they dropped, Ehrlich would pay Simon the difference.

In 1990, according to Ed Regis's account in *Wired* magazine, Ehrlich mailed Simon a cheque for $576.07. The prices of all five metals had declined, by an average of more than 50 per cent, and Ehrlich spectacularly lost his bet. Unlike Ehrlich, Simon understood that higher prices would encourage more production, and better technology would lead to declining prices over the long term. Although it is trivially true that the earth is spatially finite, it is extremely large by comparison with humanity's needs. Scarcity is a relative concept, one that is mediated by the price mechanism.

Likewise, people stubbornly refused to starve en masse. Not only has the proportion of people suffering malnourishment declined to a third of the 1970 level of 35 per cent, but the absolute number is down from over 900 million to 400 million, according to Lomborg, despite the inconvenient truth that the world's population doubled over the same period, and despite the fact that this population growth was heavily biased towards the poor world.

Thanks to the consistent and often spectacular failure of such predictions of developmental disaster, it has become clear that the earth can sustain rather more people than expected. Why is this so? Because population growth isn't relentlessly exponential after all. It doesn't simply grow until resources are depleted, at which point Gaia takes Malthusian revenge and decimates the parasite that is *Homo sapiens*.

This is, in fact, the analogy that Lovelock makes. Gaia is a living organism that strives to find a balance to support all its systems, and actively combats any threat to it. Thus terms like 'parasite', 'virus' and 'cancer' are habitually used by Gaia's cultists to describe you, me and humanity in general.

However, the global human population is likely to stabilise over time. Already its growth is slowing, both in relative and absolute terms, and the UN Development Programme estimates that the earth's total population will reach a plateau of around 9 billion people and remain stable between the years 2050 and 2300. Why is this so? Because as more people get more prosperous, life expectancy increases and mortality rates decline. As a result, people tend to have fewer children, and the children they do have contribute to this prosperity rather than detract from it.

Press a typical Gaia cultist a little, and you'll soon find agreement with the idea that overpopulation is a problem, however. And like Ehrlich, Gaiaists will soon agree to forceful measures of population control. To them, China's brutal one-child policy was a stroke of genius. Just as religious extremists look down on 'sinners' and might consider violence against non-believers to be acceptable, or even a religious duty, those who adhere to the Gaia cult sound entirely comfortable with a human culling spree. Many reflect on deadly natural disasters as a blessing in disguise and the just revenge of an overstressed planet.

Some of the more prominent population-control advocates have begun to worry about how they sound to the rest of us. John Holdren, President Barack Obama's 'science czar', was forced in 2009 to re-nounce ideas floated in two seventies-era books he co-authored with Paul and Anne Ehrlich, involving a global regime of coercive population control, including sterilisation programmes and forced abortion.

'Why is overpopulation such a radioactive topic?' pondered the left-wing magazine *Mother Jones*, before asking some overpopulation worriers to explain.[6] Among them was Ehrlich himself:

> Overpopulation, combined with overconsumption, is the ele-phant in the room. We don't talk about overpopulation because of real fears from the past – of racism, eugenics, colonialism, forced sterilization, forced family planning, plus the fears from some of contraception, abortion, and sex. We don't really talk about overconsumption because of ignorance about the eco-nomics of overpopulation and the true ecological limits of Earth.

Instead, modern Gaia followers' policies of increasingly expensive renewable energy, mounting health and safety regulations, and escalating protection for the environment also have the effect of holding the developing world down. If the cost of development is smoke-belching factories, then environmentalists want to dictate either 'green' industry or no industry at all. This is not to say that industrial development is desirable at any cost. That's clearly not true. But the Gaiaists always seem willing to sacrifice any industrialisation, development or human activity in favour of environmental protection, no matter how necessary the development is or how tenuous the environmental danger is. Many of them stand ready to use any means – vandalism, legislative force, high-seas piracy, emotional blackmail and outright lies – to achieve their antisocial aims.

But even assuming that one can accept the misanthropic notion of reducing the earth's population by a few billion and leaving the remainder poorer than they are today, such a situation would be unlikely to relieve the pressure on animal populations and environmental resources that really are deserving of protection. On the contrary: some of the worst historical environmental damage was caused on a planet peopled by only a fraction of today's population, at only a fraction of today's living standards. It was subsistence farming, wearing leather, and burning or building with wood that caused our ancestors to raze forests and hunt vast herds of meat- and leather-producing animals, many of them almost to extinction.

The evidence simply doesn't bear out the theory. With fewer people around, well-meaning do-gooders would still be fretting about some messianic mission of 'saving' the planet, as religious prophets have done for aeons. Ironically, without the spectre of over-population accompanied by images of Third World cities teeming with brown people, they'd have a harder time doing so.

The environment turns out to be pretty resilient. Though there are localised exceptions, in general the image of a fragile, super-sensitive system that could be tipped into disaster by the slightest human disturbance is simply false. Nature is robust and capable of recovering even from disasters – the worst of which are usually natural in origin.

The environment is, of course, very much worth caring about and investing in, even if only for the purely selfish reason of main-taining a productive resource base. However, one doesn't achieve this by getting hysterical about the human population and its use of natural resources. One doesn't save, say, the tiger by discredit-ing endangered-species protection with ill-conceived, politically motivated and unnecessary listings of emotional-appeal icons such as the polar bear. One doesn't achieve a better world by activist obstructionism, designed solely to limit the economic development of the world, and halt the modern world's remarkable progress towards longer, healthier and more prosperous lives for all. And one certainly doesn't earn the buy-in of other people when you tell them that the world would be better off without them.

I once encountered a new mother, who was all apologetic for having contributed to the world's population. How tragic that she couldn't conceive of her child as being anything other than a burden to humanity.

How terribly sad that our schools are teaching children not that they ought to produce more than they consume, but that their every pleasure, and even the most basic acts of living, place the planet in mortal peril.

We're raising a generation of neurotic misanthropes, because the cult of Gaia is firmly entrenched as the religious education of our schools.

WHERE FELLOW TRAVELLERS ENDED UP

There's another theory about environmentalism that rings very true. One only needs to look at the UN's frequent climate-change conferences held in exotic holiday destinations, or at the ritual observances of Earth Day, to note that there is a strong under-current of anti-capitalist sentiment in the environmental movement. Companies are routinely demonised as the enemies of the environment, and the profit motive, far from rewarding those who are able to maintain a productive, healthy and sustainable environment, is held to be a rapacious force of wholesale environmental destruction.

There's a reason for this. In the film *The Great Global Warming Swindle*, former UK Chancellor of the Exchequer Nigel Lawson and former *New Scientist* editor Nigel Calder relate the story of the miners' strikes in the late 1970s. Margaret Thatcher, who became prime minister after the unions had brought down Ted Heath's government in 1979, needed an excuse to break their power. She was a strong proponent of nuclear power, because she didn't trust union-controlled coal and Middle Eastern oil.

To her, the then-nascent theory of global warming, which had been deemed at best a fringe notion outside the scientific main-stream, was a godsend. It offered another great argument in favour of nuclear power. So, Calder relates, she went to the Royal Society with a big purse of money and told them to go away and make a case, which they did.

Climate change thus became a politically motivated theory, funded by government grants. Many new research bodies were formed, culminating in the formation of the UN Intergovernmental Panel on Climate Change.

The new theory did not only appeal to Thatcher, however. It also appealed, said Calder, to the back-to-nature environmentalist movement, which abhorred industrialisation, cars and everything that went with modern life. Of course, these myths of a bucolic past in which the noble savage was at one with nature have long existed, and Phillip Stott, professor of biogeography at the University of London, says climate change merely gave environmentalists an excuse to justify their antipathy to the United States, capitalism, consumerism and economic growth.

This would all have been small potatoes if it hadn't been for the collapse of the Soviet Union in 1989, says Patrick Moore, a co-founder of Greenpeace who left the organisation in disgust when it started campaigning to ban entire elements of the periodic table. (He tells the story himself in the film. Greenpeace, he felt, had achieved its main objective of raising awareness of environmental degradation, and had started to campaign for a worldwide ban on chlorine. Moore's response was, 'It's an element in the periodic table; I'm not sure that's in our jurisdiction.'[7]) When the Berlin Wall fell, he argues, a lot of political activists who had been sympathetic to communism were left at a loose end. They found in environmentalism a cause that allowed them to continue their anti-capitalist crusade under a green, rather than red, banner.

Several modern writers have fleshed out this idea, notably James Delingpole, a columnist for the UK *Telegraph*, in his recent book *Watermelons: The Green Movement's True Colors*. His thesis is that the environmental movement is primarily about ideology, not science. He notes the misanthropy of the overpopulation argument and the antipathy to liberty that is implicit in demanding expansive regulation and taxation to combat environmental 'crises' like climate change. Environmentalists are green on the outside, he argues, but this is merely a facade for socialism on the inside.

This view is also consistent with a favourite economic theory of environmentalists – namely, steady-state economics. A softer version of the 'de-development' that was in vogue in Ehrlich's heyday in the 1970s, it posits that the era of economic growth either is over, because of the supposed resource limits we're about to reach, or *should* be over, because humanity is rich enough and now only distribution of wealth remains as a problem.

This view is, of course, as economically unsound as it is practically false. There is a reason that we fear recessions and depressions – periods of zero or negative economic growth. They make us poorer and throw thousands of people out of paying jobs. They leave us unable even to maintain our present infrastructure and productive assets, let alone improve them.

Worse, the wholesale redistribution to solve what steady-state economists agree is the continuing problem of poverty not only requires the exercise of draconian government powers to expropriate from the 'haves' and give to the 'have-nots', but it will leave everyone poor. Redistributing all of South Africa's income equally, for example, will result in everyone earning an average income of R4 000 a month. Granted, this is more than some people earn now, but it is less than half of the limit below which one even starts paying tax. It doesn't pay the rent on a modest flat in a major city. Moreover, everyone would only earn this for a brief while, since the incentive to produce new wealth to feed humanity will have disappeared, and with it will go jobs, factories, mines and shops.

Although Delingpole writes from a First World perspective, his view of environmentalism as anti-capitalist, anti-development and fundamentally socialist in nature has much more far-reaching consequences for developing countries. Countries that have gone through periods of socialism have experienced declining prosperity and increasing poverty. Conversely, economic freedom – that is, capitalism – is strongly correlated with rising prosperity, higher levels of employment, better quality-of-life measures and lower levels of poverty.

Of course, the world isn't quite that simple. Other factors, such as war, stable institutions, such as courts, and government corruption all play a part in determining the economic progress of developing countries. Economic freedom is difficult to quantify. It manifests as a series of qualitative features of society. It requires institutions that secure private property rights and enforce contracts, the ability to freely exchange goods and capital not only within a country but also outside its borders, low barriers to entry into labour and product markets, and few controls on wages and prices.

However, no matter how you measure it, the conclusion is crystal clear. A survey of different measures of economic freedom, compared with different measures of economic prosperity, conducted by Steve Hanke and Stephen Walters for the Cato Institute, convincingly demonstrates that every major feature of economic freedom has a highly significant positive correlation with real prosperity. By contrast, the restrictive measures advocated by many environmentalists – such as outright bans on certain products or industries, or the subsidisation of favoured industries over disfavoured but more efficient alternatives – have a negative impact on prosperity.

Rich countries may be able to afford indulging in luxuries that may, for example, marginally improve air quality, but poor countries cannot. Only as they grow richer will they increasingly become able to invest in the luxury of a more pleasant, healthier environment around them. History shows that prosperous people gladly volunteer to do so.

As Moore said, by the time he left Greenpeace, most everyone had accepted its central message of sound environmental stewardship, and all it could do to continue to raise funds and remain relevant was to become ever more extreme in its positions. Today, its mission is less about stewardship and more about leaving the environment entirely untouched.

The extremist notion of humanity not as a rational actor within its environment, but as a malignant external force that ought to have no impact on nature at all, is as illogical in the rich world as it is harmful to the interests of the poor.

TAKING CARE OF THE ENVIRONMENT

For all the nostalgic appeal that environmentalists make to pre-modern humans, or to the mystical religions of the East, one idea that they reject is the attitude towards nature contained in the Judaeo-Christian tradition that is dominant in the prosperous West. The Bible calls upon humans to exercise dominion over nature, which God hands over to them as tenants, or stewards. Having replaced this precept with their own Gaia cult, it is perhaps not surprising that environmentalists reject it, but there is another, more particular reason for their rejection.

In examining the biblical notion of humanity's relationship to the environment, the obvious caveat should be made for the benefit of readers who are, like me, non-religious or who adhere to religions outside the Judaeo-Christian tradition: just because one element of a particular religious tradition has merit doesn't mean that every-thing in that religion is justified. Whether or not the rest of the Bible is of any value is a judgement best left up to the reader, but let's stipulate that few will dispute the morality of the injunctions not to steal or kill, so few will deny that there must be at least some merit in religious scripture, if only as a document of historic atti-tudes to the world's bigger problems. A similar argument no doubt goes for the religious texts of other religions. They may conflict in places with what you or I happen to believe, or with the tenets of another religion, but they also have much in common with each other and with intrinsic human morality.

As it happens, my own conclusion comes from an economic

analysis. Since people need the environment as a resource from which to live, it makes sense to protect that environment, and ensure that it remains healthy and productive. Likewise, the notion of 'dominion' over nature is not as extreme as it sounds, when interpreted to mean managing it sensibly, and protecting humanity from the dangers and risks it poses.

After all, we need to produce food, no matter what droughts and frosts the weather throws at us. We need shelter, no matter how severe the storm or cold the winter. This requires a measure of ascendancy over the forces of nature. So does the technological development that permits us to spend fewer hours working the land and more hours producing the more sophisticated products that earn us our quality of life, or more hours of leisure to enjoy what nature offers.

It does not follow that our need for economic development gives us a licence to destroy natural resources at will. This is not only economically senseless, in that it destroys the value of a capital resource, but it also harms others in any number of ways. A Catholic priest, Robert Sirico, phrases it as follows:

> Man's stewardship does not establish a license for him to destroy creation. Certainly, nature has objective value distinct from its relation to man. Nature therefore deserves our respect but not our worship, and nature must finally be seen in relation to the moral value of the human person who has the responsibility of stewardship.[8]

The Bible itself makes it clear that sound environmental stewardship does not imply a licence to destroy: 'Seemeth it a small thing unto you to have eaten up the good pasture, but ye must tread down with your feet the residue of your pastures? And to have drunk of the deep waters, but ye must foul the residue with your feet?' (Ezekiel 34:18). Conversely, however, the notion of stewardship suggests that private ownership gives people the best incentive not to exploit natural resources recklessly, but to preserve them as a sustainable source of future prosperity.

Indeed, much environmental conservation occurs on farms and nature reserves that are privately owned and managed, while a great many of the most intractable environmental problems, such as the state of fisheries or the destruction of forests, can be traced directly to the difficulty in establishing private property rights. When 'everyone' owns the resource, nobody takes responsibility for it, and the spoils go to those who are most efficient at plundering it. This observation was first made in 1968 by Garrett Hardin, who used a common pasture as an example. 'Freedom in a commons brings ruin to all,' he declared, in the essay that coined the phrase 'tragedy of the commons'.[9]

By contrast, few private farmers would be willing to increase production on their land to the point where the land is ruined. Those who do will go bankrupt, and their lands will be taken over cheaply by more responsible farmers who are better able to produce the food and other agricultural products that the market – that is, the rest of us – desires.

Likewise with the environment. There are competing interests that only a market can adjudicate. While some may desire parts of nature to remain unspoiled, this is likely to be affordable only for more attractive regions or areas of particular ecological sensitivity, where private owners are willing to invest in conservation. Other demands are equally pressing, and equally justifiable. We must eat; therefore some land has to be tilled. We must have shelter; therefore some land will be given over to housing, roads and all the infra-structure that goes with it. We must have water; so some land will be sacrificed for dams and reservoirs. It is only when the price mechanism determines the level of demand, and the most efficient way to supply that demand, that particular choices about the environment become apparent. And while it is true that prices cannot perfectly capture all information at all times, they are a reasonable approximation of the true costs and benefits that derive from a particular use of a particular piece of land.

Only when those costs and benefits are evaluated by humans who act as responsible stewards to the environment can sensible decisions be reached that adequately provide for sufficient productivity to meet the needs of society. When the right to sensible stewardship is removed, and environmentalists cast every environmental impact as an unconscionable violation of some or other natural right, or a reckless disturbance of an imagined ecological balance, we become impotent.

In a wealthy society, this is bad enough. In a country where a third of the population lives below a poverty line of R3 000 ($400) per year, and a quarter of the population remains formally unemployed, such curbs on economic development are themselves unconscionable.

Environmentalists make themselves guilty of lies and exaggeration to advance their agenda of an untouched environment free of the pernicious influence of humans they believe to be evil at heart. In doing so, they undermine their own credibility when they do raise important issues, as a result of the 'cry wolf' phenomenon. But, far more importantly, their misanthropy encourages policies that make it harder to address the very real developmental issues that poor countries face. Their actions benefit no one, except perhaps their own fund-raising departments.

The truth is that the earth is quite capable of sustaining development and growing populations. Resources, as Julian Simon illustrated, do not become relentlessly more scarce, because prices work to restrict demand and boost supply when possible. The earth is pretty large, and production isn't a zero-sum game that simply depletes resources in a one-to-one relationship with population size. What Simon understood is that the price mechanism militates against the uncontrolled exploitation that environmentalists fear, because scarcity is priced into our ability to use resources productively.

As an incidental benefit, the more prosperous we get, the more we invest in sustainability, the more sophisticated and technically skilled we get at resource management, the more we care about a healthy and productive environment, and the more we value future sustainability over present consumption.

As the rich world amply demonstrates, successful economic development is not the problem – it is the solution to the problems that population growth and resource exploitation bring. There's no reason why the same would not hold true for the developing world. Opposing development in the poor world on mistaken sustainability grounds is not only a selfish type of neocolonialism on the part of environmentalists in the rich world (or rich environmentalists in the poor world), but it is quite misanthropic. There is no social justice in perpetuating poverty by limiting economic growth and reducing economic freedom.

10

Exaggeration Makes Us Poorer

WHEN CONSERVATION CAUSES EXTINCTION

For over thirty-five years, the Convention on International Trade in Endangered Species, or CITES, has been the primary means of prohibiting what many environmentalists believe to be the underlying reason for threats to animals: the trade in products derived from them. It was recently updated to cover plants, too, and in particular rare hardwoods.

Established in 1973 and implemented in 1975 as a formalisation and expansion of earlier attempts to regulate the trade in endangered animals (such as the 1933 London Convention, which protected forty-two named species), CITES today covers nearly 35 000 species of plants and animals. It was originally ratified by only twenty-one governments, but today it is backed by 175 nations – almost all the world.

Proponents of the convention cite the fact that only one species protected by CITES – Spix's macaw – has actually gone extinct in the wild (ironically contradicting frequent claims among environmentalists of high and rising numbers of extinctions due to human pressure on the environment). Critics, however, note that the trade in endangered-species products has only risen as affluence has increased in the developing world, and criminal trade in animal, timber and fish products ranks behind only drugs, counterfeit money and human trafficking by value. At the convention's thirty-fifth anniversary dinner in 2010, Thomas Jemmi of the Swiss Federal Veterinary Office could cite only crocodilia, vicuña (a relative of the alpaca) and a few medicinal herbs as CITES success stories.

This treaty has comprehensively failed to protect, for example, the black rhino. Its population has plummeted from nearly 100 000 in 1960. According to the Red List of Threatened Species, published by the IUCN, between 1970 and 1992, large-scale poaching caused a dramatic 96 per cent collapse in numbers. By 1995, twenty years after CITES was implemented, the black rhino population reached its nadir of a mere 2 410. The result? Environmentalists insist on stricter enforcement. Plastering their advertisements with horror images of poached animals, they evoke emotion, sadness, anger and even trauma in the hope of achieving what dogged enforcement could not.

Few, if any, of those advertisements, or the news stories that result from activist press releases, will tell the truth. For example, they focus on the numbers of animals killed, which have been rising alarmingly in recent years. They focus on the astonishing price for rhino horn. They will tell you that rhino horn is prized in the Far East for its aphrodisiac properties.

These claims are only partially true, and suffer from the common tendency among some environmentalists to exaggerate or mislead the public in pursuit of apparently noble objectives. For example, they rarely note that the population growth-rate of both black and white rhinos still exceeds the rate at which animals are being poached. Populations are growing in most countries that host them, at a rate of 7.2 per cent for white rhinos and 4.8 per cent for their black counterparts (although preliminary data show that white rhino populations declined slightly for the first time in 2011). The rising level of poaching is a matter of serious concern, but it is not (yet) a genuine threat to the survival of either species. Nor do environmentalists speculate about the true reason for the sharp rise in the value of rhino horn – namely, persistent demand and severely constrained supply as a consequence of the ban in horn trade.

More damagingly, the invidious myth that rhino horn is used as an aphrodisiac continues to be promoted by many rhino conservation organisations.

The origin of this explanation goes back to the 1950s and 1960s, when colonial myths about the supposed harems and sexual habits of Chinese men were widely embellished as evidence of the godlessness, debauchery and barbarian perfidy of the inscrutable race of 'Orientals'. It was regularly repeated in conservation literature, although the lack of citations in these books suggests that they were nothing more than an uncritical repetition of what the great white hunters of yesteryear had told their credulous audiences back home.

The trouble is, there is no truth to the story.

Richard Ellis, in his excellent book on the subject, *Tiger Bone and Rhino Horn*, quotes from a range of sources, going as far back as an early translation of a 1597 Chinese treatise about the medicines then in use. He concludes: 'Even though many Westerners believe that rhino horn is used as an aphrodisiac in certain Asian countries, it is in fact used to cure almost everything *but* impotence and sexual inadequacy.'[1]

According to the environmental economist Michael 't Sas-Rolfes, rhino horn has long been prized for its ornamental value in objects such as bowls and dagger handles. The Gujarati people once believed it to have aphrodisiac properties, but this is no longer the case, and the most important demand in East Asia today relates to the medicinal properties rhino horn is believed to possess in the treatment of inflammation and fever, among other ailments.

A similar myth has arisen around tiger parts. Although one of its parts – take a wild guess which one – supposedly has aphrodisiac properties, for the most part the animal is a source of more mundane medicine.

According to the conservation organisation Tigers in Crisis: 'Endangered tiger parts such as bones, eyes, whiskers and teeth are used to treat ailments and disease ranging from insomnia and malaria, to meningitis and bad skin. Chinese texts state that the active ingredients in tiger bone, calcium and protein, which help promote healing, have anti-inflammatory properties.'[2] The 'true' uses of various tiger parts, according to Tigers in Crisis, are:

- Tiger claws: used as a sedative for insomnia
- Teeth: used to treat fever
- Fat: used to treat leprosy and rheumatism
- Nose leather: used to treat superficial wounds such as bites
- Tiger bone: used as an anti-inflammatory drug to treat rheumatism and arthritis, general weakness, headaches, stiffness or paralysis in lower back and legs, and dysentery
- Eyeballs: used to treat epilepsy and malaria
- Tail: used to treat skin diseases
- Bile: used to treat convulsions associated with meningitis in children
- Whiskers: used to treat toothaches
- Brain: used to treat laziness and pimples
- Penis: used in love potions such as tiger soup, as an aphrodisiac
- Dung or faeces: used to treat boils, haemorrhoids and cure alcoholism

The danger of simplifying the true nature of demand for products derived from endangered species is not just that the users of these products will shrug off such information campaigns as patronising interference by Western elites – it also suggests that those who are concerned about the poaching of animals for medicinal purposes do not understand the real basis for the demand, which probably won't disappear any more than the demand for homeopathic products will disappear in the West.

Moreover, the misinformation and exaggeration have serious policy consequences. For example, proposals to farm rhinos for their horns elicit very strong opposition among certain environmental groups, as well as the general public.

At the time of writing, a live rhino sells for less than the price of a kilogram of horn. The latter, in turn, is more valuable than gold. The idea behind farming rhinos is to permit the sale of rhino horn by private owners of rhino herds, which will give these animals a higher economic value. Being able to profit from them in more ways than from tourism alone will give game farmers an incentive to breed and protect their herds. At the same time, the ability to obtain horn legally – whether for decorative, medicinal or trophy purposes – will reduce the motivation to poach. Why risk getting shot poaching rhino when you can simply buy some horn?

That this kind of system works, while CITES fails, is hard to dispute. For example, in the Cayman Islands, the Cayman Turtle Farm was founded in 1968 with the goal of selling turtle meat. The farm raised turtles hatched in captivity from eggs taken from the wild, since the natural dynamics of laying eggs every year on the beaches cannot be replicated on a farm. As part of the programme, the farm repopulated the ocean by releasing year-old hatchlings back into the wild. Poachers had less incentive to hunt wild animals, because a legal market was created.

When the CITES treaty was signed, however, the major markets for turtle meat – in particular, the US – were closed to the Cayman Turtle Farm. To make matters worse, a 1979 change in the rules of the CITES convention completed the work of destroying the farm's business model. The convention used to exempt endangered animals bred in captivity from trade bans, but it now extended the ban to animals bred from eggs collected in the wild. Without viable first-generation stock, the farm collapsed.

The Cayman Turtle Farm was taken over by the island government in 1983 and supplied meat to domestic consumers. It also developed the tourist trade in an attempt to increase the commercial appeal of the sea turtles. The farm is now a shadow of its former self, but despite the sharp limitation of its market, it continues to release tens of thousands of turtles back into the sea, to grow the threatened wild population.

Farming endangered species is not unique. Brazilian farms collect wild caiman eggs, and raise the hatchlings for their meat and skin. In the case of several species, such as the Yacare caiman, which was listed as endangered in the 1980s after the indiscriminate hunting that took place in the mid-twentieth century, subsequent farming has led to a recovery in populations. Today, Yacare caiman are listed in the 'least concern' column in the IUCN Red List.

Over the past few decades, commercial ranchers have also succeeded in breeding populations of antelope species that once were on the brink of extinction. Among them are three species that are extinct in the wild: the scimitar-horned oryx, the addax and the dama gazelle. They sound like African animals, and they are. However, they're no longer found in Africa. They used to be limited to a mere handful kept caged in zoos, but several decades ago private game-farm owners began to buy surplus animals to breed them.

Today, there are 5 000 farms or ranches that belong to the Exotic Wildlife Association. Most of them are in the US state of Texas. These ranches now support large herds of oryx, addax and dama gazelles, when before there were none. In fact, Texas ranches currently support the largest populations of exotic and endangered species on earth.

The commercial sustainability of these breeding projects rests on one thing and one thing only: game hunting. Without the legal hunting of trophy animals, ranchers say, they would have neither the money nor the motive to sustain large breeding populations. Special exemptions under US endangered-wildlife legislation have created a hunting industry worth $1.3 billion, employing 14 000 people.

The CBS investigative news show *60 Minutes* recently did an insert on the surprising success of breeding near-extinct animals on hunting ranches in Texas. They spoke not only with the businessmen who made this remarkable conservation achievement a reality, but also with environmentalists like Priscilla Feral, president of Friends of Animals, an international animal rights organisation.

And Ms Feral is not amused. In fact, she is horrified by the practice. A brief extract from the show illustrates that she would rather see the animals extinct than being farmed sustainably, supported by a hunting industry. In the excerpt below, Lara Logan is the interviewer, and Charly Seale is a rancher:

FERAL: They're breeding these antelopes, they're selling the antelopes, and they're killing the antelopes. And they're calling it conserving them. They are saying it's an act of conservation and that's lunacy.

LOGAN: You would rather they did not exist in Texas at all?

FERAL: I don't want to see them on hunting ranches. I don't want to see them dismembered. I don't want to see their value in body parts. I think it's obscene. I don't think you create a life to shoot it.

LOGAN: So, if the animals exist only to be hunted …

FERAL: Right …

LOGAN: … you would rather they not exist at all?

FERAL: Not in Texas, no.

SEALE: Our biggest enemy are the animal rights people. They don't understand what we do.

LOGAN: What's to understand? I mean, you're hunters. You hunt these exotic animals. That's pretty simple.

SEALE: It is, but there are a faction of people out there that would just as soon see these animals go extinct as to have us use them for sp ... to hunt, and after all, that is the bottom line. That's what these animals are all about. That's why they are here in the numbers that they're here today.[3]

In part, the environmentalist objection is a distaste for commercial activity in general. The notion that animals are a resource that has a market value, and that this offers a reason to protect and grow their populations, is anathema to environmentalists like Feral.

Unfortunately, however, the exaggerated purity of the environmental ambition of groups like Friends of Animals has an impact on governments. In the US, Friends of Animals joined a lawsuit by the Humane Society, which petitioned the government to prohibit such hunting. The matter is still subject to litigation at the time of writing, but a prohibition seems likely.

The Exotic Wildlife Association, which represents the ranchers, describes the legislation as a 'death warrant' for the endangered animals. As Seale put it to Logan:

Just since the announcement of that rule the value of those animals has probably dropped in half. You've got to understand, I'm a rancher to make a profit, just like any business. I will say that in five years you'll see half the numbers that you see today. And I would venture to guess in ten years they'll be virtually none of 'em left.

The notion that conservation can never involve killing animals and can only involve supporting wild populations in their native habitats is not limited to hunting. Although killing animals is particularly controversial when it involves threatened species, the emotional reaction also extends to conservation-management programmes that are forced to cull animals to balance problem populations or to reduce overpopulation.

I recall as a young journalist speaking with an old game ranger during a tour of the Pilanesberg Game Reserve, near Sun City in South Africa. Surveying the landscape – this was in the mid-1990s – even a non-expert could clearly see that the park had an elephant-population problem. As far as the eye could see, trees had been trampled or had their bark removed, killing them. What trees were left of what was once thriving bushveld were little more than stunted shrubs.

'How big is your problem?' I asked him. He knew by my gesture at the dead wood that littered the barren, dusty landscape that I was referring to elephants.

'Sixty,' he replied.

Although I didn't cover environmental issues as a journalist in those days, I was a frequent visitor to South Africa's national parks, and was aware that a vigorous public campaign was being waged against culling at the time. The practice had been suspended not long before, so I asked the ranger what the park was doing to save the rest of the ecosystem from disaster.

He shrugged. 'We hunt them,' he told me, 'in the far north, where tourists don't usually notice.' However, he said the park did not have the capacity to manage more than one $100 000 hunt a month. Translocation of elephants to other game farms or reserves was problematic, not only because of the great cost of such operations, but also because there were few viable destinations that did not have overpopulation or population-imbalance problems of their own.

The shrug he gave remains in my mind to this day, as an eloquent, if silent, critique of environmentalism. Even if one assumes the best of faith on their part, the good intentions of those who profess to care about nature here paved the road to a hell of ecological destruction. The cost to South Africa's wildlife treasures is high.

Academics and conservationists continue to research effective ways of managing the ecosystems of national parks – including all forms of population control. However, many environmental lobbies and members of the public blindly refuse to countenance any methods that seem to them emotionally distressing.

Hunts soon were given the derogatory label 'canned', and became the subject of outraged newspaper headlines. So-called green hunts, in which animals are darted but not killed, were banned in 2008. A theory soon developed that culling is not acceptable on the basis that it traumatises family groups. In truth, only a single study exists that makes this claim, and research is ongoing into the implications and consequences of population management of all kinds, in the hope of finding solutions that strike a balance between being effective and affordable, on the one hand, and as humane as possible under the circumstances, on the other.

Forgotten in the emotion of anti-culling protests is that the consequences of overpopulation or imbalanced elephant populations are severe. Not only do they cause distress among the affected population itself, but the rest of the ecosystem, game-park employees, tourists and local communities all suffer greatly.

In fact, there is great irony in the fact that one of the consequences of inadequate or impotent elephant-population management is attacks on other species by problem animals. Dozens of rhino, both white and black, have been killed in Pilanesberg by elephants, long before poachers could get anywhere near them.

The perverse consequence of the emotive approach to environmentalism is often the exact opposite of what environmentalists claim to desire. Instead of protecting populations, trade bans and anti-hunting campaigns threaten species with extinction. Instead of protecting the environment, entire ecosystems get damaged by the very animals that are the photogenic subject of environmental campaign posters.

THE GREAT POLAR BEAR CRISIS

The deceptive photo of the stranded polar bears discussed in Chapter 7 was not just an isolated case of environmental exaggeration. Such imagery was a frequent accompaniment to news coverage of a major campaign by the environmental lobby.

In 2005, the Center for Biological Diversity in the US filed a petition to have the polar bear listed under the provisions of that country's Endangered Species Act (ESA). In 2008, the US Department of Fish and Wildlife complied, and declared the polar bear to be 'threatened'.

Labels such as these are not arbitrary, and the classification of a species has specific legal consequences. The ESA's definitions of 'endangered' and 'threatened' differ from those of the IUCN, which maintains the famous Red List of species that, for the most part, aren't really going extinct. By IUCN standards, a species is 'threatened' if it falls in any of the categories 'critically endangered', 'endangered' or 'vulnerable'. These terms describe an extremely high, very high or high risk of extinction, respectively.

The US, however, has different terms. It differentiates between 'endangered' species and those that are merely 'threatened'. The former is a species 'which is in danger of extinction throughout all or a significant portion of its range', while the latter merely means it is likely to become endangered in the future.

This is an important distinction, and organisations such as the International Fund for Animal Welfare pushed hard for an 'endangered' listing for polar bears, citing among its reasons that this listing offers stronger protections and would get in the way of future leases or sales of land for oil and gas drilling. It failed, but even the lower status provides an array of legal protection to a species.

The listing decision was ostensibly taken because of a threat to the polar bear's habitat, Arctic sea ice, due to climate change. The law prohibits a range of activities that threaten this habitat.

The first question that arises is whether the polar bear population really is under threat. Judging by their population numbers, this in itself seems a stretch.

Back in 2006, David Legates, director of the University of Delaware's Center for Climatic Research, conducted a study commissioned by the National Center for Policy Analysis, a non-partisan think tank dedicated to developing and promoting private, free-market alternatives to government regulation and control. He found that there was much cause for optimism about polar bear populations.

He examined several aspects of the question. About Arctic climate, he concluded that the present data were inconclusive about future sea-ice loss, and cited a Canadian Department of Fisheries and Oceans study that attributed most of the change in Arctic sea ice to wind patterns rather than temperature. In any case, he noted, polar bear populations had survived previous warm periods, about which we are aware thanks to ice cores and other historical records. It might surprise environmentalists, but animals tend to migrate in response to changes in their environment, whether those changes are caused by seasons, longer cyclical climate variation, the movement of other species, such as predators or prey, or indeed human activity.

Legates quotes the World Wide Fund for Nature, one of the organisations that lobbied for the change, to the effect that of the twenty distinct polar bear populations, accounting for some 22 000 animals, only two are declining. Two are growing, ten are stable, and the status of the remaining six is unknown. His conclusion was: 'There may be threats to the future survival of the polar bear, but global warming is not primary among them.'[4]

In a series of blog posts in 2008, I noted this study, and also analysed the available research data. I plotted the results of every population study I was able to track down. In the 1960s, there were about 10 000 polar bears, according to two studies. During the 1970s and 1980s, studies found 20 000 bears. Since the late 1990s, several studies reported numbers ranging from 20 000 to 30 000, with most agreeing to 20 000 to 25 000. In absolute numbers, then, the population has grown, or remained stable at worst.

That's not what environmental organisations wanted authorities to believe, however. Returning to the issue in 2008, the NCPA's Sterling Burnett claimed that Greenpeace and the National Resources Defense Council both cited just a single study that documented only one population in western Canada, which had suffered weight loss leading to lower cub-survival rates. He proceeded to quote a Canadian state biologist, who says that the problems in such sub-populations are most likely a result of overpopulation and competition for resources, because, in truth, that country's overall polar bear population had *grown* by 25 per cent in the previous decade.

Although the NCPA makes a strong case why the studies on which the US decision was based are flawed, their conclusions have been upheld in ongoing litigation on the subject. The reason for the litigation? Environmentalists are trying to apply the ruling to try to sue companies over their land use and greenhouse gas emissions. Not surprisingly, the courts have rejected this notion, since even if the entire theory that such emissions threaten polar bears holds true, there is no chain of evidence linking a particular company's actions to a particular case of harm.

The court rulings have upheld the listing of the polar bear as threatened, but have also upheld the principle that there can be an exemption to the ordinary requirements of the Endangered Species Act for various industrial activities.

In an editorial in 2008, *Bloomberg* writer Kevin Hassett pointed out this inconsistency in the US government's position. If it determined that carbon emissions cause global warming, which threatens Arctic sea ice, which poses a realistic threat to polar bear populations over the next fifty or one hundred years, then it cannot exempt itself from its legal obligation to enforce the provisions of the law under which that determination was made. That is, it cannot say it doesn't have to enforce limits on the emissions it claims are ultimately a threat to polar bears. If, on the other hand, it doesn't think regulating carbon emissions ought to be done via the Endangered Wildlife Act, but ought to be achieved by more specific emissions legislation, then it should not have listed a species as threatened when, in all other respects, it appears to be perfectly fine and even thriving.

The most recent rulings, from 2011, do little to resolve this conflict. For obvious reasons, environmentalists continue to challenge the exemption, and the lawsuits won't stop.

What this sorry saga of environmental politics illustrates is twofold. The first is that the extinction or otherwise of the polar bear is not the primary issue for environmentalists. In fact, they do not even seem to care whether it really is threatened, as their selective use of data demonstrates. The second is that the fierce polar bear has replaced the docile panda as the mascot for the environmental movement because what environmentalists are really after is a legal club with general application. Unlike the panda, the claimed threat to polar bears can be used against a wide range of industrial concerns even if they are far removed from polar bear habitats, if environmentalists believe them to be responsible not for extinctions per se, but for climate change.

THE ECONOMIC HARM OF EXAGGERATION

There is surprisingly little research about the economic impact of environmental regulation. Conflicting arguments exist. On the one hand, the business sector says, with much anecdotal evidence to show for it, that regulation strangles productivity and growth. On the other hand, environmentalists argue that green technology offers growth prospects, and in any case, so-called externalities are not counted in ordinary financial reports.

Externalities are those costs, impacts or indeed benefits that are not accounted for in financial terms by the organisation engaging in production. This is an intuitively appealing way to consider the implications of environmental effects.

Positive effects – such as fertile land, high biodiversity, sustainable forests, vaccination or clean water – clearly do have a value to people who did not pay directly for those benefits. Likewise, negative effects – such as pollution – cause harm to those affected by them, without the costs being borne by the polluter. Much regulation is based on this principle, although it often founders on the problem that externalities are hard to measure, and their causes hard to prove.

However, in considering externalities, one ought to consider more than just environmental benefits. A healthy environment is clearly of benefit to those who require its resources for the production of food or a place to live. So is the infrastructure that comes with industrial development. Roads and other public infrastructure are often laid on for industrial concerns, and by means of rates and taxes, these companies also pay for them. When a city grows around factories or harbours, the benefits to the city of the economic activity and infrastructure generated by the companies responsible are incidental benefits that are hard to measure.

When considering negative externalities, it is reasonable to engage with companies to determine to what extent costs might be borne by people other than their shareholders, employees and customers. When these can clearly be determined, existing law is capable of assigning a value to this externality, and requiring that compensation be paid.

On the other hand, exaggerating the effects of so-called externalities is not justifiable. Undertaking endless research studies to try to place an accurate value on them merely serves to postpone economic activity, and ultimately reduces the value of this activity, or increases its cost.

A similar argument holds for regulatory cost-benefit decisions. Without knowing both the costs and the benefits, policymakers are at a loss whether or not to permit a given activity, or how to assign costs and liabilities.

When the latest version of a set of recommendations for good business practice, commissioned by South Africa's Institute of Directors – the King III report on corporate governance – was launched, I attended the presentation by its author, retired justice Mervyn King. He held that 'risk management' for companies that are responsibly run ought to extend beyond mere financial risk, and should include social and environmental risks. As an example, he warned that sea levels might rise by eight metres in the next 100 years, and wanted to know if the assembled managers in the audience had plans for this contingency.

Now one must forgive King. He is not an expert on environmentalism. And this kind of exaggeration is routinely drummed into us by scientists, a massive environmental lobby with vested interests in keeping climate alarmism alive, and a credulous media that thrives on sensationalist tales of doom.

But evaluating risk means working with correct data. And what are the correct data in this case? If you accept the IPCC predictions – and there are many reasons to be sceptical of those – sea levels could rise by anything between nine centimetres and eighty-eight centimetres by 2100, mostly as a result of the heat expansion of ocean water.

That's a tremendously large margin of error, and anyone schooled in the finer points of risk management would find it difficult to assess a proper response. Other predictions centre on a modest thirty-centimetre rise.

This is, of course, nothing out of the ordinary. Sea levels rose by almost that much during the twentieth century, too. Yet one doesn't often hear old people reminisce about the world wars, the moon landing, the release of Nelson Mandela, the atom bomb, the advent of cars and computers, and add, 'Oh yes, and sea levels rose by a foot. That was scary.'

The number King used was twenty-five times higher than the consensus forecast. It was ten times higher than the IPCC's worst-case scenario. Even the perpetual climate doom-monger James Hansen, the head of NASA's Goddard Institute for Space Science, which maintains one of the primary global-climate records on which the IPCC bases its political conclusions, can't reach to much more than half of King's prediction. Heaven alone knows where King got that number, but a prudent risk manager in command of all the relevant data would conclude that the chances of it happening are as near zero as doesn't make a difference.

If he'd been serious about risk management, King would have correctly quantified the extent of the catastrophe he warned about, and assessed its likelihood, which is extremely low. He would also have compared the relative cost of mitigation – preventing sea levels from rising, starting today – and adaptation – adding a couple of feet to sea walls and flood defences in fifty years' time.

Exaggerating the likely extent of environmental harm makes such evaluations difficult or impossible. As a consequence, there are two distinct dangers involving public policymakers and private individuals, and both are particularly harmful in the context of relatively poor, developing economies.

The one possibility is that valid warnings will be ignored as yet another case of environmental exaggeration. As this book demonstrates time and again, crying wolf is a routine (and apparently deliberate) part of the media and fund-raising strategy of environmental organisations.

The other is the opposite danger: that policy decisions on the part of private persons or public officials will be based on exaggerated estimates of likely harm. The cost of over-regulation can have severe consequences for the companies on which citizens rely for jobs, food, and other goods and services. It can also delay or thwart economic development that could improve the outlook for prosperity in developing countries.

The energy crisis in South Africa provides a case in point. The country is in the grip of a serious supply constraint as a consequence of a lack of investment in generation and distribution capacity in the last twenty years. New projects are on the cards. Some large coal-fired power stations have already been commissioned, but other alternatives are also on the table.

Among them is the development of nuclear power stations, which might cost anywhere between R300 billion and R1 trillion.

Policymakers and the general public need to establish which of these numbers is true, who stands to benefit from the build contracts, and how to address the ordinary safety concerns that go along with nuclear power.

In doing so, it is perfectly reasonable to learn from what happened at Fukushima. But it is not reasonable to parade a Japanese farmer through the local townships, as Greenpeace has been doing, to whip up a groundswell of support for their mission: that South Africa must 'abandon its nuclear expansion plans in favour of a strong push to energy efficiency and renewable power'. One cannot draw statistically valid conclusions form isolated anecdotes. In South Africa, the geological circumstances are not the same, the reactor designs won't be forty years old and tsunamis aren't likely. And even if they were, Fukushima has yet to result in a single human death. By contrast, coal mines kill thirty people a day, and even solar panels, hydroelectricity and biofuels have a more deadly safety record than nuclear power.

The shale-gas controversy, too, is a case in point. The South African government instituted a moratorium on issuing exploration licences, pending a study to determine whether the opposition to hydraulic fracturing has any merit. At the time of writing, this moratorium remains in place, despite the fact that very little other than anecdotal tales and weakly supported claims of risk have been uncovered in the intervening year.

Meanwhile, oil and gas companies are sitting on their hands, busying themselves with developments elsewhere in the world. Besides losing the early-mover advantage to rivals in shale-gas production such as the US, China and Poland, there are other economic benefits that South Africa is, at best, postponing as a result of the environmentalists' delay tactics.

Worse, if the campaign to prohibit shale-gas exploitation is successful, none of those economic benefits will be realised. When doing a risk-benefit analysis, it is not sufficient to assess only the risks and to maximise them 'just in case'. Policymakers – who in South Africa own the mineral rights, so are the only ones in a position to permit or deny shale-gas drilling – need to know the true risks, on the one hand, and the likely benefits, on the other.

Economist Tony Twine, mere weeks before his untimely death in 2012, offered a summary of the potential benefits that might accrue to South Africa if the government's study group returns a favourable verdict on shale gas. The Karoo shale holds a vast gas resource of 485 trillion cubic feet. It is currently estimated to be the fifth largest in the world and compares extremely favourably with the 1 trillion cubic feet on which the Mossgas project was justified.

Over the period when it is likely to be productive, until 2035, South Africa's demand for electricity will likely triple, assuming a sustained annual GDP growth rate of 4.5 per cent. (At a lower growth level of 3 per cent, demand will have almost doubled, and if the country achieves 7 per cent growth, electricity demand will rise sixfold.)

South Africa is already in the grip of a serious electricity-supply crunch, which has forced the national energy utility to periodically institute rolling blackouts and demand that major industrial users scale back their electricity consumption. The only reason the last few years have not been a lot worse is because the global economic downturn has put a damper on South Africa's own growth and reduced its industrial electricity demand. Deliberately opposing energy production seems a perverse response to the reality that low growth and low energy demand go hand in hand.

Using traditional economic modelling techniques, involving the so-called Keynesian multiplier effect (the validity of which is not only beyond the scope of this book, but is also not very relevant to the argument), Twine assesses the economic impacts of shale-gas production in this light. He considers two conservative scenarios: the production of 20 trillion cubic feet over twenty-five years, and the production of 50 trillion cubic feet over the same time span. The annual value this would add to the South African economy ranges between R80 billion and R200 billion. Counting direct and indirect employment, he expects anywhere between 300 000 and 700 000 jobs to be generated.

There are over 4.2 million unemployed adults in South Africa. Another 2.3 million are discouraged work-seekers. That represents 24 per cent and 37 per cent of the total labour force of 17.7 million, respectively. The labour force in turn represents 54 per cent of the people between the ages of fifteen and sixty-four, and 35 per cent of the total population. Employing 700 000 people, if all else remains equal, would mean providing financial support for a population of 2 million. In a developing country like South Africa, such a tremendous boost to the economy and to poverty alleviation cannot be lightly dismissed on the basis of environmental scaremongering. In fact, even a very much lower economic return would merit exploiting the shale-gas reserves South Africa possesses.

At the time of writing, the Treasure the Karoo Action Group had promised to take the government to court if its study came to the conclusion that shale-gas exploration and eventual production ought to be permitted.

It is reasonable to be cautious and seek the best possible information needed to develop an economy while also mitigating environmental risks. However, a developing country can ill afford unreasonable delays and costs in property development as a result of exaggerated fears raised during the environmental-impact assessment phase. A poor population cannot afford products that are more expensive because of irrational fears over public health. A sluggish growth rate and high unemployment are not remedied when entire industries can be thwarted as a result of politically motivated scare-mongering.

Those who are rich enough to be able to tour the country, taking pretty photographs and stoking up environmental fears, may be able to afford to indulge their emotional exaggeration. Those who live on the breadline, however, cannot. This is why the Karoo Shale Gas Community Forum, an umbrella body that represents the interests of various youth organisations, small farmers' associations, churches, community groups and labour unions, have not only come out in favour of permitting shale-gas exploration, but have actively distanced themselves from the Treasure the Karoo Action Group. They know that opportunities such as these are few and far between in a developing country where millions earn very little, and millions more are unemployed.

There are serious policy questions that need to be answered about matters that may affect the environment, such as shale gas and nuclear power. Such questions can only be clouded by environmental exaggeration. The consequences can only harm the fragile growth of an emerging economy in a tough global climate.

Endnotes

Chapter 1

1 Lewis Pugh's official website, 'Standing up to Goliath', http://www. lewispugh.com.

2 Maarten J. de Wit, 'The great shale debate in the Karoo', *South African Journal of Science* 107, no. 7/8 (2011), art. 791, http://www.sajs.co.za/ index.php/SAJS/article/view/791/730.

Chapter 2

1 Quotations in this section are taken from Colorado Oil and Gas Conservation Commission, '*Gasland*', press release, http://cogcc.state. co.us/library/GASLAND%20DOC.pdf.

2 Phelim McAleer, '*Gasland* director hides full facts', Not Evil Just Wrong, 1 June 2011, http://www.noteviljustwrong.com/General/ gasland-director-hides-full-facts.html.

3 Joshua Kors, 'Oscar nominee Josh Fox speaks out about oil lobby's efforts to crush his film', *Huffington Post*, 27 January 2011, http://www. huffingtonpost.com/joshua-kors/director-josh-fox-receive_b_814590. html.

4 Ivo Vegter, 'Karoo fracking scandal exposed!', *Daily Maverick*, 13 April 2011, http://dailymaverick.co.za/opinionista/2011-04-13- karoo-fracking-scandal-exposed.

5   Agency for Toxic Stubstances and Disease Registry, 'ToxFAQs™ for 2-butoxyethanol and 2-butoxyethanol acetate', August 1999, http://www.atsdr.cdc.gov/toxfaqs/tf.asp?id=346&tid=61.

6   Andrew Maykuth, '"Gasland" documentary fuels debate over natural gas extraction', *Philadelphia Inquirer*, 23 June 2010, http://articles. philly.com/2010-06-23/news/24961785_1_natural-gas-marcellus- shale-gas-drilling.

7   Bob Niedbala, 'Tributary in trouble', *Observer-Reporter*, 6 March 2010, http://www.observer-reporter.com/or/story11/06-03-2010-mon- river-makes-list.

8   Abrahm Lustgarten, 'Scientific study links flammable drinking water to fracking', *ProPublica*, 9 May 2011, http://www.propublica.org/ article/scientific-study-links-flammable-drinking-water-to-fracking.

9   Stephen Osborn et al., 'Methane contamination of drinking water accompanying gas-well drilling and hydraulic fracturing', *Proceedings of the National Academy of Sciences* 108, no. 20 (17 May 2011): 8172– 8176, http://www.pnas.org/content/108/20/8172.full.

10 Chevron, 'Chevron opens largest wastewater treatment plant in SA', press release, 22 August 2008, http://www.petroleumafrica.com/en/ newsarticle.php?NewsID=6345.

11 UK House of Commons Energy and Climate Change Committee, 'Shale gas: Fifth report of Session 2010–12', December 2010, http://www.publications.parliament.uk/pa/cm201012/cmselect/ cmenergy/795/795.pdf.

Chapter 3

1   Maarten J. de Wit, 'The great shale debate in the Karoo', *South African Journal of Science* 107, no. 7/8 (2011), art. 791, http://www.sajs.co.za/ index.php/SAJS/article/view/791/730.

2   Melanie Gosling, 'Shell to recycle fracking water', *Cape Times*, 23 May 2011, http://www.iol.co.za/scitech/science/environment/ shell-to-recycle-fracking-water-1.1072392.

3   UK House of Commons Energy and Climate Change Committee, 'Shale gas: Fifth report of Session 2010–12', December 2010,

http://www.publications.parliament.uk/pa/cm201012/cmselect/
cmenergy/795/795.pdf.

4   Stacy Shackford, 'Natural gas from fracking could be "dirtier" than
    coal, Cornell professors find', *Chronicle Online*, Cornell University,
    11 April 2011, http://www.news.cornell.edu/stories/April11/
    GasDrillingDirtier.html.

5   Ben Geman, 'EPA official calls Cornell gas-climate study "important
    piece of information"', *The Hill*, 12 April 2011, http://thehill.com/
    blogs/e2-wire/e2-wire/155503-epa-official-calls-cornell-gas-
    climate-study-important-piece-of-information.

6   Michael Levi, 'Some thoughts on the Howarth shale gas paper',
    Council on Foreign Relations, 15 April 2011, http://blogs.cfr.org/
    levi/2011/04/15/some-thoughts-on-the-howarth-shale-gas-
    paper/.

7   David McCabe, 'Let's fix dangerous, climate-warming methane leaks
    from all fossil fuels: Coal, oil, and natural gas', Clean Air Task Force,
    13 April 2011, http://www.catf.us/blogs/ahead/2011/04/13/lets-fix-
    dangerous-climate-warming-methane-leaks-from-all-fossil-fuels-
    coal-oil-and-natural-gas/.

8   Nathan Hultman et al., 'The greenhouse impact of unconventional
    gas for electricity generation', *Environmental Research Letters* 6, no. 4
    (25 October 2011), http://iopscience.iop.org/1748-9326/6/4/044008.

9   Steve Everley, 'Shale gas: Cornell's GHG paper continues to attract
    criticism', *MasterResource*, 2 November 2011, http://www.
    masterresource.org/2011/11/shale-gas-cornell-criticism/.

10 *Ibid.*

11 David McCabe, 'Let's fix dangerous, climate-warming methane leaks
    from all fossil fuels: Coal, oil, and natural gas', Clean Air Task Force,
    13 April 2011, http://www.catf.us/blogs/ahead/2011/04/13/lets-fix-
    dangerous-climate-warming-methane-leaks-from-all-fossil-fuels-
    coal-oil-and-natural-gas/.

12 John Hanger, 'Gas is cleaner than coal: What Cornell prof gets right
    and wrong', 12 April 2011, http://johnhanger.blogspot.com/2011/04/
    gas-was-and-is-cleaner-than-coal.html.

13 Ivo Vegter, 'A fracture in SA's astronomical advantage?', *ITWeb Brainstorm*, 7 June 2011, http://www.brainstormmag.co.za/index.php?option=com_content&view=article&id=4228:a-fracture-in-sas-astronomical-advantage.

14 'South Africa spotlight on earthquake', *AON Benfield*, June 2010, http://www.aon.com/attachments/reinsurance/201006_mega_eq_report.pdf.

15 Siseko Njobeni, 'Karoo's shale gas to boost economy, says report', *Business Day*, 2 March 2011, http://www.businessday.co.za/articles/Content.aspx?id=166487.

16 Treasure the Karoo Action Group, official website, 'About us', http://treasurethekaroo.co.za/.

17 Max Matavire, 'Fracking gets support', *The New Age*, 1 July 2011, http://www.thenewage.co.za/21979-1016-53-Fracking_gets_support.

Chapter 4

1 Quotations are from Rachel Carson, *Silent Spring* (Boston: Houghton Mifflin Harcourt, 2002), pp. 1–2.

2 Paul Reiter, 'Dangers of disinformation', *International Herald Tribune*, 11 January 2007, http://www.nytimes.com/2007/01/11/opinion/11iht-edreiter.4171294.html.

3 National Network for Immunization Information, 'History and achievements', 2010, http://www.immunizationinfo.org/parents/why-immunize/history-and-achievements.

4 Shauna Wood, 'What is in a vaccine???' *Vaccination News*, 9 August 2000, http://www.vaccinationnews.com/dailynews/may2001/whatsinvax.htm.

5 This and the following quotation are from Bill Ahearn, 'The autism-vaccines myth: The impact of the media', *Psychology Today*, 8 February 2010.

6 See Agence France-Presse, 'Autism researcher a "victim of smear campaign"', as published by ABC Australia, 7 January 2011, http://www.abc.net.au/news/2011-01-06/autism-researcher-a-victim-of-smear-campaign/1896750.

7 'Vaccines: The case of measles', *Nature* 473 (23 May 2011): 434–435, doi:10.1038/473434a, http://www.nature.com/news/2011/110525/full/473434a.html.

8 'The birth of Aji-No-Moto', corporate website, http://www.ajinomoto.com/features/aji-no-moto/en/basic/index.html.

9 Katherine Zeratsky, 'Monosodium glutamate (MSG): Is it harmful?', Mayo Clinic, http://www.mayoclinic.com/health/monosodium-glutamate/an01251.

10 European Food Information Council, 'The facts on monosodium glutamate', November 2002, http://www.eufic.org/article/en/artid/monosodium-glutamate/.

11 A. Rosalie David and Michael R. Zimmerman, 'Cancer: An old disease, a new disease or something in between?' *Nature Reviews Cancer* 10 (October 2010), doi:10.1038/nrc2914, http://www.nature.com/nrc/journal/v10/n10/abs/nrc2914.html.

12 Bjørn Lomborg, 'A roadmap for the planet', *Newsweek*, 12 June 2011, http://www.thedailybeast.com/newsweek/2011/06/12/bjorn-lomborg-explains-how-to-save-the-planet.html.

13 George Johnson, 'Unearthing prehistoric tumors, and debate', *New York Times*, 27 December 2010, http://www.nytimes.com/2010/12/28/health/28cancer.html.

Chapter 5

1 Ivo Vegter, 'I'm ashamed for my profession', *Daily Maverick*, 4 April 2011, http://dailymaverick.co.za/opinionista/2011-04-04-im-ashamed-for-my-profession.

2 Matthew Wald and David Jolly, 'Dangerous levels of radioactive isotope found 25 miles from nuclear plant', *New York Times*, 31 March 2011, http://www.nytimes.com/2011/03/31/world/asia/31japan.html.

3 International Atomic Energy Agency, *IAEA International Fact-Finding Expert Mission of the Nuclear Accident Following the Great East Japan Earthquake and Tsunami*, 1 June 2011, http://www.iaea.org/newscenter/focus/fukushima/missionsummary010611.pdf.

4   Bibi van der Zee, 'Japan nuclear crisis puts UK public off new power stations', *Guardian*, 22 March 2011, http://www.guardian.co.uk/environment/2011/mar/22/japan-nuclear-crisis-uk-power-stations.

5   John Williams, 'Save Bantamsklip', *Save Bantamsklip Association* website, http://www.savebantamsklip.org/.

6   Neil Shaw (producer) and Annika Larsen (presenter), 'Nuke', M-Net *Carte Blanche*, 27 June 2010, http://beta.mnet.co.za/carteblanche/Article.aspx?Id=4012&ShowId=1.

7   Christopher Busby Foundation for the Children of Fukushima, official website, http://www.cbfcf.org/.

8   'Radiation risk from nuclear power station in Japan', *BBC News*, 14 March 2011, http://www.dailymotion.com/video/xhlgks_radiation-risk-from-nuclear-power-station-in-japan_news.

9   Reuters, 'Alarm over plutonium', 29 March 2011, http://www.timeslive.co.za/world/article994217.ece/Alarm-over-plutonium.

10 MIT Dept of Nuclear Science and Engineering, 'Plutonium in the environment', misdated 18 March 2011, http://mitnse.com/2011/03/18/323/.

11 Melanie Gosling, 'Koeberg sitting alongside fault line', *Cape Times*, 16 March 2011, http://www.iol.co.za/scitech/science/environment/koeberg-sitting-alongside-fault-line-1.1042362.

12 Bibi van der Zee, 'Japan nuclear crisis puts UK public off new power stations', *Guardian*, 22 March 2011, http://www.guardian.co.uk/environment/2011/mar/22/japan-nuclear-crisis-uk-power-stations.

13 Catherine Brahic, 'Nuked coral reef bounces back', *New Scientist*, 14 April 2008, http://www.newscientist.com/article/dn13668.

14 Ivo Vegter, *TEDxJohannesburg* live blog, 15 November 2009, http://ivo.co.za/2009/11/15/tedxjohannesburg-live-blog/.

15 Mara Hvistendahl, 'Coal ash is more radioactive than nuclear waste', *Scientific American*, 13 December 2007, http://www.scientificamerican.com/article.cfm?id=coal-ash-is-more-radioactive-than-nuclear-waste.

16 'Radiation risk from nuclear power station in Japan', *BBC News*, 14 March 2011, http://www.dailymotion.com/video/xhlgks_radiation-risk-from-nuclear-power-station-in-japan_news.

17 Agence France-Presse, 'Fukushima much bigger than Chernobyl
    – expert', 2 April 2011, http://www.news24.com/World/News/
    Fukushima-much-bigger-than-Chernobyl-expert-20110402.

18 'Nuclear radiation "the greatest public health hazard"', CNN,
    25 March 2011, http://edition.cnn.com/2011/OPINION/03/25/
    caldicott.nuclear.health/index.html.

19 '*New York Times* quietly edits article about Fukushima evacuation',
    Japan Probe, 25 March 2011, http://www.japanprobe.com/2011/03/26/
    new-york-times-quietly-edits-article-about-fukushima-evacuation/.

20 Hiroko Tabuchi, Keith Bradsher and David Jolly, 'Japan encourages a
    wider evacuation from reactor area', *New York Times*, 25 March 2011,
    http://www.nytimes.com/2011/03/26/world/asia/26japan.html.

21 George Monbiot, 'Why Fukushima made me stop worrying and love
    nuclear power', *Guardian*, 21 March 2011, http://www.guardian.co.
    uk/commentisfree/2011/mar/21/pro-nuclear-japan-fukushima.

22 Paul Joseph Watson, 'Nuclear expert: Monbiot "criminally
    irresponsible" for downplaying Fukushima', *Infowars*, 28 March 2011,
    http://www.infowars.com/nuclear-expert-monbiot-criminally-
    irresponsible-for-downplaying-fukushima/.

Chapter 6

1 Naomi Klein, 'Gulf oil spill: A hole in the world', *Guardian*, 19 June
   2010, http://www.guardian.co.uk/theguardian/2010/jun/19/
   naomi-klein-gulf-oil-spill.

2 Larissa MacFarquhar, 'Outside agitator: Naomi Klein and the new
   new left', *New Yorker*, 8 December 2008, http://www.newyorker.com/
   reporting/2008/12/08/081208fa_fact_macfarquhar.

3 Terry Macalister, 'Tony Hayward: How an affable geologist became
   America's most hated', *Guardian*, 9 June 2010, http://www.guardian.
   co.uk/business/2010/jun/09/tony-hayward-bp-barack-obama.

4 Michael Graham Richard, 'Gulf of Mexico oil spill: The what, when
   and where', *PlanetGreen*, 3 June 2010, http://planetgreen.discovery.
   com/tech-transport/gulfofmexico-oilspill-whatwhenwhere-
   whatyoucando.html.

5   Michael Graham Richard, 'The "Katrina of smell" is attacking
    News Orleans thanks to BP oil spill', *Treehugger*, 30 April 2010,
    http://www.treehugger.com/corporate-responsibility/the-katrina-of-
    smell-is-attacking-news-orleans-thanks-to-bp-oil-spill.html.

6   Pablo Paster, 'Ask Pablo: Why didn't they just burn the BP oil spill?',
    Treehugger, 3 May 2010, http://www.treehugger.com/natural-
    sciences/ask-pablo-why-didnt-they-just-burn-the-bp-oil-spill.html.

7   Brian Merchant, 'Is the Gulf oil spill really that big of a deal?',
    Treehugger, 4 May 2010, http://www.treehugger.com/corporate-
    responsibility/is-the-gulf-oil-spill-really-emthatem-big-of-a-
    deal.html.

8   Vern Radul, 'Deepwater Horizon: The Gulf Stream is only the
    beginning', *Antemedius*, 5 May 2010, http://antemedius.com/content/
    deepwater-horizon-gulf-stream-only-beginning.

9   Christine Dell'Amore, 'Gulf oil spill could reach east coast beaches',
    National Geographic, 4 May 2010, http://news.nationalgeographic.
    com/news/2010/05/100504-science-environment-gulf-oil-spill-
    loop-current-florida/.

10 David Biello, 'Where will the Deepwater Horizon oil end up?',
    Scientific American, 19 May 2010, http://www.scientificamerican.com/
    article.cfm?id=where-will-the-deepwater-horizon-oil-end-up.

11 Vern Radul, 'Deepwater Horizon: The Gulf Stream is only the
    beginning', *Antemedius*, 5 May 2010, http://antemedius.com/content/
    deepwater-horizon-gulf-stream-only-beginning.

12 Jill Schneiderman, 'Pandora's oil well', *EarthDharma*, 11 May 2010,
    http://jillschneiderman.wordpress.com/2010/05/11/pandoras-well/.

13 National Commission on the BP Deepwater Horizon Oil Spill
    and Offshore Drilling, *Final Report to the President: Deepwater:
    The Gulf Oil Disaster and the Future of Offshore Drilling*, January 2011,
    http://www.oilspillcommission.gov/sites/default/files/documents/
    DEEPWATER_ReporttothePresident_FINAL.pdf.

14 National Wildlife Federation, 'How does the BP oil spill impact
    wildlife and habitat?', http://www.nwf.org/Oil-Spill/Effects-on-
    Wildlife.aspx.

15 Philip Sherwell, 'BP oil spill: Dramatic recovery of Gulf of Mexico one year on', *Telegraph*, 10 April 2011, http://www.telegraph.co.uk/finance/newsbysector/energy/oilandgas/8423173/BP-oil-spill-Dramatic-recovery-of-Gulf-of-Mexico-one-year-on.html.

16 Christopher Reddy, 'How science failed during the Gulf oil disaster', *Wired*, as reprinted in *Ars Technica*, 22 April 2012, http://arstechnica.com/science/news/2012/04/how-science-failed-during-the-gulf-oil-disaster.ars.

17 Michael Grunwald, 'The BP spill: Has the damage been exaggerated?', *Time*, 29 July 2010, http://www.time.com/time/magazine/article/0,9171,2007428,00.html.

18 United Nations Environment Programme, *Updated Scientific Report on the Environmental Effects of the Conflict between Iraq and Kuwait*, 10–21 May 1993, www.unep.org/dewa/westasia/data/Knowledge_Bases/Iraq/Reports/UNEPGCIraq1993.pdf.

19 'Scientists find that tons of oil seep into the Gulf of Mexico each year', *Science Daily*, 27 January 2000, http://www.sciencedaily.com/releases/2000/01/000127082228.htm.

20 Keith Clarke and Jeffrey Hemphill, 'The Santa Barbara oil spill: A retrospective', University of California, Santa Barbara, http://www.geog.ucsb.edu/~kclarke/Papers/SBOilSpill1969.pdf.

21 Greenpeace, 'BP Deepwater disaster and Gulf oil spill', 15 June 2010, http://www.greenpeace.org/usa/en/news-and-blogs/news/gulf-oil-spill/#a3.

Chapter 7

1 Stephen Hawking, *A Brief History of Time: From the Big Bang to Black Holes* (New York: Bantam, 1988).

2 Power Balance company website, 'Power Balance is performance technology', 2011, http://www.powerbalance.com/powerbalance.

3 Bhaskar Prasad, 'Power Balance bracelet company files for bankruptcy', *International Business Times*, 21 November 2011, http://www.ibtimes.com/articles/253724/20111121/power-balance-bracelet-company-files-bankruptcy.htm.

4   Andreas Späth, 'Much ado about fracking', News24, 20 April 2011, http://www.news24.com/Columnists/AndreasSpath/Much-ado-about-fracking-20110420.

5   Rachel Adatia, 'Anti-fracking activists call for a national boycott of Shell!', *Earthlife Africa*, 16 May 2011, http://www.earthlife.org.za/?p=1591.

6   'What goes in and out of hydraulic fracturing', *Dangers of Fracking* website, http://www.dangersoffracking.com/.

7   Bill Mouland, 'Global warming sees polar bears stranded on melting ice', *Daily Mail*, 1 February 2007, http://www.dailymail.co.uk/news/article-433170/Global-warming-sees-polar-bears-stranded-melting-ice.html.

8   'Gore pays for photo after Canada didn't', *National Post*, 23 March 2007, http://www.canada.com/nationalpost/news/story.html?id=5961259b-de08-4532-850b-09d4753bed39.

9   Elizabeth Rosenthal and Andrew Revkin, 'Science panel calls global warming "unequivocal"', *New York Times*, 3 February 2007, http://www.nytimes.com/2007/02/03/science/earth/03climate.html.

10 *Ibid.*

11 'Gore pays for photo after Canada didn't', *National Post*, 23 March 2007, http://www.canada.com/nationalpost/news/story.html?id=5961259b-de08-4532-850b-09d4753bed39.

12 Greenpeace, 'The Advanced Energy [R]evolution', 24 May 2011, http://www.greenpeace.org/africa/en/News/news/The-Advanced-Energy-Revolution-Report/.

13 Ebrahim Patel, 'The New Growth Path: The framework', South African Government Information, 23 November 2010, http://www.info.gov.za/aboutgovt/programmes/new-growth-path/index.html.

14 Greenpeace, 'The Advanced Energy [R]evolution', 24 May 2011, http://www.energyblueprint.info/fileadmin/media/documents/national/2011/E_R__South_Africa_May_2011-LR.pdf.

15 Karla Zadnik et al., 'Vision: Myopia and ambient night-time lighting', *Nature*, 9 March 2000, http://www.nature.com/nature/journal/v404/n6774/abs/404143a0.html.

16 'A severe strain on credulity', *New York Times*, 13 January 1920, http://web.archive.org/web/20070217065558/http://it.is.rice. edu/~rickr/goddard.editorial.html.

17 'A correction', *New York Times*, 17 July 1969. A copy is archived at: Bjorn Carey, 'New York Times to NASA: You're right, rockets do work in space', *Popular Science*, 20 July 2009, http://www.popsci.com/ military-aviation-amp-space/article/2009-07/new-york-times-nasa-youre-right-rockets-do-work-space.

18 The Nizkor Project website, http://www.nizkor.org/.

19 'Eisenhower and the righteous cause: The liberation of Europe', *Eisenhower Foundation* website, 2010, http://www.dwightdeisenhower .com/pdfs/righteous_cause_prospectus.pdf.

20 Peter Christoff, 'Climate change is another grim tale to be treated with respect', *The Age*, 9 July 2007, http://www.theage.com.au/news/ opinion/climate-change-is-another-grim-tale-to-be-treated-with-res pect/2007/07/08/1183833338608.html.

Chapter 8

1 A. P. Møller et al., 'Dispersal and climate change: A case study of the Arctic tern *Sterna paradisaea*', *Global Change Biology*, October 2006, http://onlinelibrary.wiley.comdoi/10.1111/j.1365-2486.2006.01216.x/ full.

2 Shanta Barley, 'Arctic tern crowned "king of commuters"', *New Scientist*, 13 January 2010, http://www.newscientist.com/article/ dn18379-arctic-tern-crowned-king-of-commuters.html.

3 Martin Durkin (producer and director), *The Great Global Warming Swindle*, documentary film premiered on 8 March 2007, Channel 4, UK, http://www.youtube.com/watch?v=YtevF4B4RtQ.

4 'History of Europe', *Britannica Online Encyclopaedia*, http://www. britannica.com/EBchecked/topic/195896/history-of-Europe/276190/Demographic-and-agricultural-growth.

5 'Little Ice Age (LIA)', *Britannica Online Encyclopaedia*, http://www. britannica.com/EBchecked/topic/344106/Little-Ice-Age-LIA.

6 Ross McKitrick, 'What is the "hockey stick" debate about?',

University of Guelph, 4 April 2005, http://www.uoguelph.
ca/~rmckitri/research/McKitrick-hockeystick.pdf.

7   Richard A. Muller, 'A prime piece of evidence linking human
    activity to climate change turns out to be an artifact of poor
    mathematics', *MIT Technology Review*, 15 October 2004,
    http://www.technologyreview.com/energy/13830/.

8   Ivo Vegter, 'Pop goes the hot air balloon', *Daily Maverick*,
    24 November 2009, http://dailymaverick.co.za/opinionista/2009-
    11-24-Pop-goes-the-hot-air-balloon.

9   Ivo Vegter, 'Climate fraud kills people', *Daily Maverick*, 1 December
    2009, http://dailymaverick.co.za/opinionista/2009-12-01-Climate-
    fraud-kills-people.

10 Andrew Montford, 'Climate cuttings 33', *Bishop Hill*, 20 November
    2009, http://bishophill.squarespace.com/blog/2009/11/20/climate-
    cuttings-33.html.

11 Andrew Montford, 'The Yamal deception', *Bishop Hill*, 9 May 2012,
    http://www.bishop-hill.net/blog/2012/5/9/the-yamal-deception.
    html.

12 'READ ME for Harry's work on the CRU TS2.1/3.0 datasets,
    2006-2009!', http://www.anenglishmanscastle.com/HARRY_READ_
    ME.txt.

13 Eric S. Raymond, 'Hiding the decline: Part 1 – The adventure begins',
    Armed and Dangerous, 24 November 2009, http://esr.ibiblio.
    org/?p=1447.

14 Anthony Watts, 'Climategate 2 email – Rob Wilson replicates
    McIntyre & McKitrick – produces hockey sticks out of noise',
    Watts Up With That?, 27 November 2011, http://wattsupwiththat.
    com/2011/11/27/climategate-2-email-briffa-replicates-mcintyre-
    mckitrick-produces-hockey-sticks-out-of-noise/.

15 Steve McIntyre, 'Briffa on another Mann hockey stick', *Climate Audit*,
    22 November 2009, http://climateaudit.org/2009/11/22/briffa-on-
    another-mann-hockey-stick/.

16 Steve Milloy, 'Climategate 2.0: Jones, Briffa say Mann, hockey stick
    "on dodgy ground"', *Junk Science*, 28 November 2011, http://

junkscience.com/2011/11/28/climategate-2-0-jones-briffa-say-
mann-hokey-stick-on-dodgy-ground/.

17 Richard A. Muller, 'A prime piece of evidence linking human activity
to climate change turns out to be an artifact of poor mathematics',
MIT Technology Review, 15 October 2004, http://www.
technologyreview.com/energy/13830/.

18 Roger Pielke Jr, 'GRL and James Saiers', blog post, 28 November
2011, http://rogerpielkejr.blogspot.com/2009/11/grl-and-james-
saiers.html.

19 David Adam and Fred Pearce, 'No apology from IPCC chief
Rajendra Pachauri for glacier fallacy', *Guardian*, 2 February 2010,
http://www.guardian.co.uk/environment/2010/feb/02/climate-
change-pachauri-un-glaciers.

20 Ben Webster, 'Climate chief was told of false glacier claims before
Copenhagen', *The Times*, 30 January 2010, http://www.timesonline.
co.uk/tol/news/environment/article7009081.ece.

21 Christopher Booker, 'African crops yield another catastrophe for the
IPCC', *Telegraph*, 13 February 2010, http://www.telegraph.co.uk/
comment/columnists/christopherbooker/7231386/African-crops-
yield-another-catastrophe-for-the-IPCC.html.

22 '"I feel duped on climate change"', *Der Spiegel*, 8 February 2012,
http://www.spiegel.de/international/world/breaking-global-
warming-taboos-i-feel-duped-on-climate-change-a-813814.html.

23 Anthony Watts, 'Zorita calls for barring Phil Jones, Michael Mann,
and Stefan Rahmstorf from further IPCC participation', *Watts Up
With That?*, 27 November 2009, http://wattsupwiththat.
com/2009/11/27/zorita-calls-for-barring-phil-jones-michael-mann-
and-stefan-rahmstorf-from-further-ipcc-participation/.

24 Andrew Revkin, 'A climate scientist who engages skeptics', *New York
Times*, 27 November 2009, http://dotearth.blogs.nytimes.
com/2009/11/27/a-climate-scientist-on-climate-skeptics/.

25 William Alexander, 'Climate change: May it rest in peace', 19 May
2010, http://climaterealists.com/attachments/database/2010/
Climate%20Change%20May%20It%20Rest%20In%20Peace.pdf.

26 S.I. Rasool and S.H. Schneider, 'Atmospheric carbon dioxide and aerosols: Effects of large increases on global climate', *Science*, 9 July 1971, http://www.sciencemag.org/content/173/3992/138.abstract.

27 Wallace Broecker, 'Climatic change: Are we on the brink of a pronounced global warming?', *Science*, 8 August 1975, http://www.sciencemag.org/content/189/4201/460.abstract.

28 Stephen Schneider, 'Don't bet all environmental changes will be beneficial', *American Physical Society News*, August/September 1996, http://www.americanphysicalsociety.com/publications/apsnews/199608/upload/aug96.pdf.

29 Oliver Morton, 'Heroes of the environment', *Time*, 17 October 2007, http://www.time.com/time/specials/2007/article/0,28804,1663317_1663319_1669893,00.html.

30 James Lovelock, 'The earth is about to catch a morbid fever that may last as long as 100,000 years', *Independent*, 16 January 2006, http://www.independent.co.uk/opinion/commentators/james-lovelock-the-earth-is-about-to-catch-a-morbid-fever-that-may-last-as-long-as-100000-years-523161.html.

31 Sarah Sands, 'We're all doomed! 40 years from global catastrophe – and there's NOTHING we can do about it, says climate change expert', *Daily Mail*, 22 March 2008, http://www.dailymail.co.uk/news/article-541748/Were-doomed-40-years-global-catastrophe--theres-NOTHING-says-climate-change-expert.html.

32 Leo Hickman, 'James Lovelock: Humans are too stupid to prevent climate change', *Guardian*, 29 March 2010, http://www.guardian.co.uk/science/2010/mar/29/james-lovelock-climate-change.

33 Ian Johnston, '"Gaia" scientist James Lovelock: I was "alarmist" about climate change', MSNBC, 23 April 2012, http://worldnews.msnbc.msn.com/_news/2012/04/23/11144098-gaia-scientist-james-lovelock-i-was-alarmist-about-climate-change.

34 Ivo Vegter, 'Ten reasons to reject climate alarmism', *Daily Maverick*, 2 March 2010, http://dailymaverick.co.za/opinionista/2010-03-02-ten-reasons-to-reject-climate-alarmism.

35 James Hansen, 'Game over for the climate', *International Herald Tribune*, 9 May 2012, http://www.nytimes.com/2012/05/10/opinion/game-over-for-the-climate.html.

36 David Fahrenthold, 'CFC replacements intensify climate concerns', *Washington Post*, 20 July 2009, http://www.washingtonpost.com/wp-dyn/content/article/2009/07/19/AR2009071901817.html.

37 'Plenty of gloom', *The Economist*, 18 December 1997, http://www.economist.com/node/455855.

Chapter 9

1   Michael Crichton, 'Environmentalism as religion', 15 September 2003, http://scienceandpublicpolicy.org/commentaries_essays/crichton_three_speeches.html.

2   James Lovelock, 'What is Gaia?', undated, http://www.ecolo.org/lovelock/what_is_Gaia.html.

3   Paul Ehrlich, 'Getting at the roots of terrorism', Stanford News Service, 15 November 2002, http://news.stanford.edu/pr/02/ehrlichiraq1120.html.

4   Bjørn Lomborg, *The Skeptical Environmentalist: Measuring the Real State of the World* (Cambridge: Cambridge University Press, 2001), pp. 252ff.

5   Julian Simon, 'The statistical flummery of species loss'. 13 October 1993, http://www.juliansimon.com/writings/Articles/SPECIFOR.txt.

6   Julia Whitty, 'Why is population control such a radioactive topic?', Mother Jones, 12 May 2010, http://www.motherjones.com/blue-marble/2010/05/population-forum.

7   Martin Durkin (producer and director), *The Great Global Warming Swindle*, documentary film premiered on 8 March 2007, Channel 4, UK, http://www.youtube.com/watch?v=YtevF4B4RtQ.

8   Robert Sirico, 'Receiving the gift of environmental stewardship', Acton Institute for the Study of Religion and Liberty, 3 May 2000, http://conservation.catholic.org/gift_of_stewardship.htm.

9   Garrett Hardin, 'The tragedy of the commons', *Science*, 13 December 1968, www.sciencemag.org/cgi/reprint/162/3859/1243.pdf.

Chapter 10

1   Richard Ellis, *Tiger Bone and Rhino Horn: The Destruction of Wildlife for Traditional Chinese medicine* (Washington, DC: Island Press, 2005).

2   'Traditional Chinese medicine and tigers', Tigers in Crisis website, http://www.tigersincrisis.com/traditional_medicine.htm.

3   Max McClellan (producer) and Lara Logan (correspondent), 'Can hunting endangered animals save the species?' CBS *60 Minutes*, 29 January 2012, http://www.cbsnews.com/8301-18560_162-57368000/can-hunting-endangered-animals-save-the-species/.

4   H. Sterling Burnett, 'Polar bears on thin ice, not really!' National Center for Policy Analysis, 17 May 2006, http://www.ncpa.org/pub/ba551.

REFERENCES

BOOKS

Binfield, Kevin (ed.). *Writings of the Luddites*. Baltimore and London: Johns
  Hopkins University Press, 2004

Booker, Christopher. *The Real Global Warming Disaster: Is the Obsession with
  'Climate Change' Turning Out to be the Most Costly Scientific Blunder in History?*
  London and New York: Continuum International Publishing Group, 2009

Carson, Rachel. *Silent Spring*. Boston: Houghton Mifflin Harcourt, 2002

Copi, Irving, et al. *Introduction to Logic*. Upper Saddle River, N.J: Prentice
  Hall, 2010

Delingpole, James. *Watermelons: How the Environmentalists are Killing the Planet,
  Destroying the Economy and Stealing Your Children's Future*. London: Biteback
  Publishing, 2012

———. *Watermelons: The Green Movement's True Colors*. New York: Publius
  Books, 2011

Ehrlich, Paul. *The Population Bomb*. New York: Buccaneer Books, 1997

Eisenbud, Merrill, and Tom Gesell. *Environmental Radioactivity from Natural,
  Industrial and Military Sources*. San Diego: Academic Press, 1997

Ellis, Richard. *Tiger Bone and Rhino Horn: The Destruction of Wildlife for
  Traditional Chinese Medicine*. Washington, D.C: Island Press, 2005

Hawking, Stephen. *A Brief History of Time: From the Big Bang to Black Holes*.
  New York: Bantam, 1988

International Union for Conservation of Nature and Natural Resources.
  Crocodiles: Proceedings of the 6th Working Meeting of the Crocodile Specialist

Group of the Species Survival Commission of the International Union for Conservation of Nature and Natural Resources, September 1982. Gland, Switzerland: IUCN, 1984

Klein, Naomi. *No Logo.* New York: Picador, 2002

Lawson, Nigel. *An Appeal to Reason: A Cool Look at Global Warming.* London and New York: Duckworth Overlook, 2009

Lomborg, Bjørn. *Cool It: The Skeptical Environmentalist's Guide to Global Warming.* Singapore: Marshall Cavendish, 2009

————. *The Skeptical Environmentalist: Measuring the Real State of the World.* Cambridge: Cambridge University Press, 2001

Meadows, Donella, et al. *Beyond the Limits.* White River Junction, V.T: Chelsea Green Publishing Company, 1992

————. *The Limits to Growth: A Report to the Club of Rome's Project on the Predicament of Mankind.* New York: Universe Books, 1972

Montford, A.W. *The Hockey Stick Illusion: Climategate and the Corruption of Science.* London: Stacey International, 2010

Myers, M.G., and D. Pineda. 'Misinformation about vaccines', in D.T. Barrett and L.R. Stanberry (eds), *Vaccines for Biodefense and Emerging and Neglected Diseases.* London: Elsevier, 2009

Myers, Norman. *The Sinking Ark: A New Look at the Problem of Disappearing Species.* New York: Pergamon Press, 1979

National Research Council. *Oil in the Sea III: Inputs, Fates, and Effects.* Washington, D.C: National Academies Press, 2003

Nilsson, Greta. *Endangered Species Handbook.* Washington, D.C: Animal Welfare Institute, 1983 <http://www.endangeredspecieshandbook.org/>

Oreskes, Naomi, and Erik Conway. *Merchants of Doubt.* New York: Bloomsbury, 2010

SA Department of Water Affairs. *The Groundwater Dictionary* (2nd ed.). Published online 2011 <http://www.dwaf.gov.za/Groundwater/Groundwater_Dictionary/index.html>

Scholes, R.J., and K.G. Mennell (eds). *Elephant Management: A Scientific Assessment for South Africa.* Johannesburg: Witwatersrand University Press, 2008

Sinclair, Matthew. *Let Them Eat Carbon: The Price of Failing Climate Change Policies and How Governments and Big Business Profit from Them.* London: Biteback Publishing, 2011

Subramanya, K. *Engineering Hydrology.* New Delhi: Tata McGraw-Hill, 2008

Walton, Douglas. *Media Argumentation: Dialectic, Persuasion, and Rhetoric*.
Cambridge: Cambridge University Press, 2007

Weston, Anthony. *A Rulebook for Arguments*. Indianapolis: Hackett, 2000

ARTICLES AND BLOG POSTS

'1000+ Peer-reviewed papers supporting skeptic arguments against
ACC/AGW alarm'. *Popular Technology* [online resource], 17 May 2012
<http://www.populartechnology.net/2009/10/peer-reviewed-papers-
supporting.html>

Adam, David, and Fred Pearce. 'No apology from IPCC chief Rajendra
Pachauri for glacier fallacy'. *Guardian* [online edition], 2 February 2010
<http://www.guardian.co.uk/environment/2010/feb/02/climate-
change-pachauri-un-glaciers>

Aerts, Jeroen, and Wouter Botzen. 'Climate change impacts on pricing
long-term flood insurance: A comprehensive study for the Netherlands'.
Global Environmental Change 21 (3), 2011, pp. 1045–1060 <http://dx.doi.
org/10.1016/j.gloenvcha.2011.04.005>

AFP. 'Activists want polar bear on endangered list before Alaska oil sale'.
4 February 2008 <http://afp.google.com/article/
ALeqM5hcin6YhX0sChAERuWHqGLSpbAHjw>

———. 'Autism researcher a "victim of smear campaign"'. *ABC News*
Australia [online], 7 January 2011 <http://www.abc.net.au/news/2011-
01-06/autism-researcher-a-victim-of-smear-campaign/1896750>

———. 'Fukushima much bigger than Chernobyl – expert'. *News24* [online
news site], 2 April 2011 <http://www.news24.com/World/News/
Fukushima-much-bigger-than-Chernobyl-expert-20110402>

Ahearn, Bill. 'Dishonest, discredited, and absent: Wakefield is thoughtless at
home'. *Psychology Today*, 31 January 2010

———. 'The autism-vaccines myth: The impact of the media'.
Psychology Today, 8 February 2010

Alexander, William. 'Climate change: May it rest in peace'. *Climate Realists*
[online blog], 19 May 2010 <http://climaterealists.com/attachments/
database/2010/Climate%20Change%20May%20It%20Rest%20In%20
Peace.pdf>

'A severe strain on credulity'. *New York Times* [on web.archive.org archive],
13 January 1920 <http://web.archive.org/web/20070217065558/http://
it.is.rice.edu/~rickr/goddard.editorial.html>

Barley, Shanta. 'Arctic tern crowned "king of commuters"'. *New Scientist*
[online edition], 13 January 2010 <www.newscientist.com/article/
dn18379-arctic-tern-crowned-king-of-commuters.html>

Bate, Roger. 'DDT works'. *Prospect* [online edition], 24 May 2008 <http://
www.prospectmagazine.co.uk/2008/05/ddtworks/>

Bennett, Geraldine. 'Karoo series: Shell's SA stakeholders uncovered'.
Moneyweb [online investment information], 13 April 2011 <http://www.
moneyweb.co.za/mw/view/mw/en/page295046?oid=535221&sn=2009
+Detail&pid=294690>

Biello, David. 'Where will the Deepwater Horizon oil end up?' *Scientific
American* [online edition], 19 May 2010 <http://www.scientificamerican.
com/article.cfm?id=where-will-the-deepwater-horizon-oil-end-up>

Blaine, Sue. 'A Karoo community's views on fracking'. *Business Day* [online
blog], 1 February 2012 <http://blogs.businessday.co.za/sue/2012/02/01/
a-karoo-communitys-views-on-fracking/>

Block, Walter. 'Coase and Demsetz on private property rights'. *Journal of
Libertarian Studies* 2 (1), 1977, pp. 111–115 <http://mises.org/journals/
jls/1_2/1_2_4.pdf>

Booker, Christopher. 'African crops yield another catastrophe for the IPCC'.
The Telegraph [online news site], 13 February 2010 <http://www.
telegraph.co.uk/comment/columnists/christopherbooker/7231386/
African-crops-yield-another-catastrophe-for-the-IPCC.html>

'BP Deepwater disaster and Gulf oil spill'. *Greenpeace* [online blog], 15 June
2010 <http://www.greenpeace.org/usa/en/news-and-blogs/news/
gulf-oil-spill/>

Brahic, Catherine. 'Nuked coral reef bounces back'. *New Scientist* [online
edition], 14 April 2008 <http://www.newscientist.com/article/dn13668>

Brayton, Ed. 'Fracking chemicals found in Wyoming groundwater'. *The
Colorado Independent* [online news site], 12 November 2011 <http://
coloradoindependent.com/105803/fracking-chemicals-found-in-
wyoming-groundwater>

Broecker, Wallace S. 'Climatic change: Are we on the brink of a pronounced
global warming?' *Science* 189 (4201), 8 August 1975, pp. 460–463 <http://
www.sciencemag.org/content/189/4201/460.abstract>

Brooks, Michael. '13 things that do not make sense'. *New Scientist* 2491,
19 March 2005, pp. 30–38 <http://www.newscientist.com/article/
mg18524911.600-13-things-that-do-not-make-sense.html>

Bryson, Donna. 'Anti-nuclear Japanese farmer visits South Africa'. Associated Press [on yahoo.com online news], 29 February 2012 <http://news.yahoo.com/anti-nuclear-japanese-farmer-visits-south-africa-182950022.html>

Burnett, H. Sterling. 'Polar bears on thin ice, not really!' *National Center for Policy Analysis* [online resource], 17 May 2006 <http://www.ncpa.org/pub/ba551>

Busby, Christopher. 'Calcium and other supplements to protect against internal radiation'. Uploaded to *Scribd* [online resource], 7 November 2011 <http://www.scribd.com/doc/72048273/supplementrept>

Carey, Bjorn. 'New York Times to NASA: You're right, rockets DO work in space'. *Popular Science* [online edition], 20 July 2009 <http://www.popsci.com/military-aviation-amp-space/article/2009-07/new-york-times-nasa-youre-right-rockets-do-work-space>

Carte, David. 'Karoo gas could give SA huge boost'. *Moneyweb* [online investment information], 26 May 2011 <www.moneyweb.co.za/mw/view/mw/en/page295023?oid=537852&sn=2009+Detail&pid=287226>

Casselman, Ben. 'Rig owner had rising tally of accidents'. *Wall Street Journal* [online edition], 10 May 2010 <online.wsj.com/article/SB10001424052748704307804575234471807539054.html>

'Chernobyl radiation killed nearly one million people: New book'. *Environment News Service* [online news site], 26 April 2010 <http://www.ens-newswire.com/ens/apr2010/2010-04-26-01.html>

Christoff, Peter. 'Climate change is another grim tale to be treated with respect'. *The Age* [online edition], 9 July 2007 <http://www.theage.com.au/news/opinion/climate-change-is-another-grim-tale-to-be-treated-with-respect/2007/07/08/1183833338608.html>

CITES Secretariat. 'A brief history of CITES'. *CITES World* Special Edition, 3 March 2003, pp. 2–3 <http://www.cites.org/eng/news/world/30special.pdf>

Clarke, Keith C., and Jeffrey J. Hemphill. 'The Santa Barbara oil spill: A retrospective'. Darrick Danta (ed.), *Yearbook of the Association of Pacific Coast Geographers* 64, 2002, pp. 157–162 <http://www.geog.ucsb.edu/~kclarke/Papers/SBOilSpill1969.pdf>

Creamer, Martin. 'Now Anglo applies to explore for shale gas in Karoo – Petroleum Agency SA'. *Mining Weekly*, 26 March 2010 <http://www.miningweekly.com/article/now-anglo-applies-to-explore-for-shale-gas-in-karoo-petroleum-agency-sa-2010-03-26>

Crichton, Michael. 'Environmentalism as religion – 15 September 2003'. *Three Speeches by Michael Crichton, SPPI Essays and Commentary Series* [PDF], 9 December 2009, pp. 14–20 <http://scienceandpublicpolicy. org/images/stories/papers/commentaries/crichton_3.pdf>

Cumming, D.H.M., et al. 'Elephants, woodlands and biodiversity in southern Africa'. *South African Journal of Science* 93, 1997, pp. 231–236 <http:// libres.uncg.edu/ir/uncg/f/M_Kalcounis-Ruppell_Elephants_1997.pdf>

Curry, Judith. 'Hiding the decline'. *Climate Etc.* [online blog], 22 February 2011 <http://judithcurry.com/2011/02/22/hiding-the-decline/>

David, A. Rosalie, and Michael R. Zimmerman. 'Cancer: An old disease, a new disease or something in between?' *Nature Reviews Cancer* 10, October 2010, pp. 728–733 <http://www.nature.com/nrc/journal/v10/n10/abs/ nrc2914.html>

Dell'Amore, Christine. 'Gulf oil spill could reach east coast beaches'. *National Geographic* [Daily News online], 4 May 2010 <http://news. nationalgeographic.com/news/2010/05/100504-science-environment- gulf-oil-spill-loop-current-florida/>

De Morsella, Chris. 'British Petroleum's Deepwater Horizon oil spill in the Loop Current – Florida next'. *The Green Economy Post* [online blog], May 2010 <http://greeneconomypost.com/bp-oil-spill-loop-current- florida-10134.htm>

De Wit, Maarten J. 'The great shale debate in the Karoo'. *South African Journal of Science* 107, (7/8), 2011, article 791 <http://www.sajs.co.za/ index.php/SAJS/article/view/791/730>

Edwards, Tim. 'Monbiot joins Lovelock in the nuclear power camp'. *The Week* [online edition], 22 March 2011 <http://www.theweek.co.uk/ people-news/6855/monbiot-joins-lovelock-nuclear-power-camp>

Efstathiou Jr, Jim, and Katarzyna Klimasinska. 'U.S. to slash Marcellus Shale gas estimate 80%'. *Bloomberg* [online news], 23 August 2011 <http:// www.bloomberg.com/news/2011-08-23/u-s-to-slash-marcellus-shale- gas-estimate-80-.html>

Ehrlich, Paul. 'Getting at the roots of terrorism'. *Stanford News Service* [online News Release], 15 November 2002 <http://news.stanford.edu/pr/02/ ehrlichiraq1120.html>

European Food Information Council. 'The facts on monosodium glutamate'. *Food Today* 35, November 2002 <http://www.eufic.org/article/en/artid/ monosodium-glutamate/>

Eustace, Michael. 'Rhino poaching: What is the solution?' *Business Day* [online edition], 20 January 2012 <http://www.businessday.co.za/Articles/Content.aspx?id=162979>

Everley, Steve. 'Shale gas: Cornell's GHG paper continues to attract criticism'. *MasterResource* [online blog], 2 November 2011 <http://www.masterresource.org/2011/11/shale-gas-cornell-criticism/>

Fahrenthold, David. 'CFC replacements intensify climate concerns'. *Washington Post* [online edition], 20 July 2009 <http://www.washingtonpost.com/wp-dyn/content/article/2009/07/19/AR2009071901817.html>

Fang, Ferric, and Arturo Casadevall. 'Retracted science and the retraction index'. *IAI Accepts* [electronic version], 8 August 2011 <http://iai.asm.org/cgi/reprint/IAI.05661-11v1.pdf>

Fowler, Tom. 'Criminal charges are prepared in BP spill'. *Wall Street Journal* [online edition], 29 December 2011 <http://online.wsj.com/article/SB10001424052970203899504577126871591624572.html>

'Gas patch scientists explain how hydraulic fracturing can permanently contaminate public water supplies'. *The Checks and Balances Project* [online blog], 6 May 2011 <http://checksandbalancesproject.org/2011/05/06/gas-patch-scientists-explain-how-hydraulic-fracturing-can-permanently-contaminate-public-water-supplies/>

Geman, Ben. 'EPA official calls Cornell gas-climate study "important piece of information"'. *The Hill* [E² Wire online blog], 12 April 2011 <http://thehill.com/blogs/e2-wire/e2-wire/155503-epa-official-calls-cornell-gas-climate-study-important-piece-of-information>

Goodman, David. 'Updates on the earthquake and tsunami in Japan'. *New York Times* [online edition], 11 March 2011 <http://thelede.blogs.nytimes.com/2011/03/11/video-of-the-earthquake-and-tsunami-in-japan/>

'Gore pays for photo after Canada didn't'. *National Post* [on Canada.com news site], 23 March 2007 <http://www.canada.com/nationalpost/news/story.html?id=5961259b-de08-4532-850b-09d4753bed39>

Gorski, David. 'Toxic myths about vaccines'. *Science-Based Medicine* [online], 18 February 2008 <http://www.sciencebasedmedicine.org/index.php/toxic-myths-about-vaccines/>

Gosling, Melanie. 'Koeberg sitting alongside fault line'. *Cape Times*, 16 March 2011 <http://www.iol.co.za/scitech/science/environment/koeberg-sitting-alongside-fault-line-1.1042362>

————. 'Shell "to recycle fracking water"'. *Cape Times*, 23 May 2011 <http://www.iol.co.za/scitech/science/environment/shell-to-recycle-fracking-water-1.1072392>

Grunwald, Michael. 'The BP spill: Has the damage been exaggerated?' *TIME* magazine [*TIME* online], 29 July 2010 <http://www.time.com/time/magazine/article/0,9171,2007428,00.html>

Hanger, John. 'Gas is cleaner than coal: What Cornell prof gets right and wrong'. *John Hanger's Facts of the Day* [online blog], 12 April 2011 <http://johnhanger.blogspot.com/2011/04/gas-was-and-is-cleaner-than-coal.html>

Hanke, Steve, and Stephen Walters. 'Economic freedom, prosperity, and equality: A survey'. *Cato Journal* 17 (2), 1997, pp. 117–146 <http://www.cato.org/pubs/journal/cj17n2-1.html>

Hansen, James. 'Game over for the climate'. *New York Times* [online edition], 9 May 2012 <http://www.nytimes.com/2012/05/10/opinion/game-over-for-the-climate.html>

Hardin, Garrett. 'The tragedy of the commons'. *Science* 162 (3859), 13 December 1968, pp. 1243–1248 <www.sciencemag.org/cgi/reprint/162/3859/1243.pdf>

Hassett, Kevin. 'Polar bear ruling to bring tsunami of lawsuits'. *Bloomberg* [online news], 19 May 2008 <http://www.bloomberg.com/apps/news?pid=20601039&refer=columnist_hassett&sid=apMzVf6jrh94>

Hickman, Leo. 'James Lovelock: Humans are too stupid to prevent climate change'. *Guardian* [online edition], 29 March 2010 <http://www.guardian.co.uk/science/2010/mar/29/james-lovelock-climate-change>

'History of Europe'. *Encyclopaedia Britannica* [online] <http://www.britannica.com/EBchecked/topic/195896/history-of-Europe/276190/Demographic-and-agricultural-growth>

Holden, Constance. 'Spilled oil looks worse on TV'. *Science* 250 (4979), 19 October 1990, p. 371

Horton, Robert. 'Court vacates polar bear special rule, upholds ban on importation of sport-hunted trophies'. *Endangered Species Law and Policy* [online resource], 18 October 2011 <http://www.endangeredspecieslawandpolicy.com/2011/10/articles/litigation/court-vacates-polar-bear-special-rule-upholds-ban-on-importation-of-sporthunted-trophies/>

Howarth, Robert W., et al. 'Methane and the greenhouse-gas footprint of natural gas from shale formations'. *Climatic Change Letters* 106 (4), March 2011, pp. 679–690 <http://www.eeb.cornell.edu/howarth/ Howarth%20et%20al%20%202011.pdf>

'How does the BP oil spill impact wildlife and habitat?' *National Wildlife Federation* [online resource], 2010 <http://www.nwf.org/Oil-Spill/ Effects-on-Wildlife.aspx>

'How the media covered the Gulf oil spill disaster'. *Pew Research Center* [online publication], 25 August 2010 <http://pewresearch.org/ pubs/1707/media-coverage-analysis-gulf-oil-spill-disaster>

Hultman, Nathan, et al. 'The greenhouse impact of unconventional gas for electricity generation'. *Environmental Research Letters* 6 (4), 25 October 2011 <http://iopscience.iop.org/1748-9326/6/4/044008>

Hvistendahl, Mara. 'Coal ash is more radioactive than nuclear waste'. *Scientific American* [online edition], 13 December 2007 <http:// www.scientificamerican.com/article.cfm?id=coal-ash-is-more- radioactive-than-nuclear-waste>

'"I feel duped on climate change"'. *Der Spiegel* [translation on Spiegel Online International], 8 February 2012 <http://www.spiegel.de/international/ world/breaking-global-warming-taboos-i-feel-duped-on-climate- change-a-813814.html>

'Japan earthquake: Tsunami hits north-east'. *BBC News Asia-Pacific* [online], 11 March 2011 <http://www.bbc.co.uk/news/world-asia- pacific-12709598>

'Japan nuke plant water "leaking into sea"'. *Sky News* [online], 2 April 2011 <news.sky.com/skynews/Home/World-News/Japan-Fukushima- Nuclear-Power-Plant-Radioactive-Water-Leaking-From-Cracks-In- Reactor-Concrete/Article/201104115964569>

Jiang, Mohan, et al. 'Life cycle greenhouse gas emissions of Marcellus shale gas'. *Environmental Research Letters* 6 (3), 5 August 2011 <http:// iopscience.iop.org/1748-9326/6/3/034014>

Johnson, George. 'Unearthing prehistoric tumors, and debate'. *New York Times*, 28 December 2010 <http://www.nytimes.com/2010/12/28/ health/28cancer.html>

Johnston, Ian. '"Gaia" scientist James Lovelock: I was "alarmist" about climate change'. *msnbc.com* [online news site], 23 April 2012 <http:// worldnews.msnbc.msn.com/_news/2012/04/23/11144098-gaia-scientist- james-lovelock-i-was-alarmist-about-climate-change>

King, Bill. 'Hysterical media reports exaggerate extent of spill'. *Houston Chronicle* [online news site], 18 August 2010 <http://www.chron.com/opinion/outlook/article/King-Hysterical-media-reports-8232-exaggerate-1713398.php>

Kirkland, Joel, and Climatewire. 'Natural gas could serve as "bridge" fuel to low-carbon future'. *Scientific American* [online edition], 25 June 2010 <http://www.scientificamerican.com/article.cfm?id=natural-gas-could-serve-as-bridge-fuel-to-low-carbon-future>

Klein, Naomi. 'Gulf oil spill: A hole in the world'. *Guardian* [online edition], 19 June 2010 <http://www.guardian.co.uk/theguardian/2010/jun/19/naomi-klein-gulf-oil-spill>

Kors, Joshua. 'Oscar nominee Josh Fox speaks out about oil lobby's efforts to crush his film'. *Huffington Post* [online news site], 27 January 2011 <http://www.huffingtonpost.com/joshua-kors/director-josh-fox-receive_b_814590.html>

Kwok, Robert Ho Man. 'Chinese restaurant syndrome'. *New England Journal of Medicine* 18 (178),1968, p. 796

Lawrence, Mark. 'Natural gas: Should fracking stop? (Robert Howarth and Anthony Ingraffea debate Terry Engelder)'. *Atkinson Center for a Sustainable Future* [online blog], 21 September 2011 <http://blog.acsf.cornell.edu/?p=804>

Leibbrandt, M., et al. 'Trends in South African income distribution and poverty since the fall of apartheid'. *OECD Social, Employment and Migration Working Papers* No. 101 [electronic download], OECD Publishing, 2010 <http://www.npconline.co.za/MediaLib/Downloads/Home/Tabs/Diagnostic/Economy2/Trends%20in%20South%20African%20Income%20Distribution%20and%20Poverty%20since%20the%20Fall%20of%20Apartheid.pdf>

Levi, Michael. 'Some thoughts on the Howarth shale gas paper'. *Council on Foreign Relations* [online blog], 15 April 2011 <http://blogs.cfr.org/levi/2011/04/15/some-thoughts-on-the-howarth-shale-gas-paper/>

'Little Ice Age (LIA)'. *Encyclopaedia Britannica* [online] <http://www.britannica.com/EBchecked/topic/344106/Little-Ice-Age-LIA.>

Lockitch, Keith. 'Rachel Carson's genocide'. *Capitalism Magazine* [online magazine], 23 May 2007 <http://www.capitalismmagazine.com/science/environment/pollution/4965-rachel-carson-s-genocide.html>

Lomborg, Bjørn. 'A roadmap for the planet'. *Newsweek* [online edition],

12 June 2011 <http://www.thedailybeast.com/newsweek/2011/06/12/
bjorn-lomborg-explains-how-to-save-the-planet.html>

Lovelock, James. 'The earth is about to catch a morbid fever that may last as
long as 100,000 years'. *Independent* [online edition], 16 January 2006
<http://www.independent.co.uk/opinion/commentators/james-
lovelock-the-earth-is-about-to-catch-a-morbid-fever-that-may-last-as-
long-as-100000-years-523161.html>

Lustgarten, Abrahm.'EPA finds compound used in fracking in Wyoming
aquifer'. *ProPublica* [online news site], 10 November 2011 <http://
www.propublica.org/article/epa-finds-fracking-compound-in-wyoming-
aquifer>

————. 'Scientific study links flammable drinking water to fracking'.
ProPublica [online news site], 9 May 2011 <http://www.propublica.org/
article/scientific-study-links-flammable-drinking-water-to-fracking>

Lyons, Rob. 'The bear necessities of climate change politics'. *spiked* [online
blog], 16 March 2007 <www.spiked-online.com/index.php?/site/
article/2969/>

Macalister, Terry. 'Tony Hayward: How an affable geologist became
America's most hated'. *Guardian* [online edition], 9 June 2010
<http://www.guardian.co.uk/business/2010/jun/09/tony-hayward-
bp-barack-obama>

Macfarquhar, Larissa. 'Outside agitator: Naomi Klein and the new new left'.
The New Yorker [online edition], 8 December 2008 <http://www.
newyorker.com/reporting/2008/12/08/081208fa_fact_macfarquhar>

Matavire, Max. 'Fracking gets support'. *The New Age* [online], 1 July 2011
<http://www.thenewage.co.za/21979-1016-53-Fracking_gets_support>

Matthiessen, Peter. 'Environmentalist Rachel Carson'. *TIME* magazine
[*TIME* online], 29 March 1999 <http://www.time.com/time/magazine/
article/0,9171,990622,00.html>

Maykuth, Andrew. '*Gasland* documentary fuels debate over natural gas
extraction'. *The Philadelphia Inquirer,* 23 June 2010 <http://articles.philly.
com/2010-06-23/news/24961785_1_natural-gas-marcellus-shale-gas-
drilling>

McAleer, Phelim. '*Gasland* director hides full facts'. *Not Evil Just Wrong*
[online blog], 1 June 2011 <http://www.noteviljustwrong.com/General/
gasland-director-hides-full-facts.html>

McCabe, David. 'Let's fix dangerous, climate-warming methane leaks from all fossil fuels: Coal, oil, and natural gas'. *Clean Air Task Force* [online blog], 13 April 2011 <http://www.catf.us/blogs/ahead/2011/04/13/lets-fix-dangerous-climate-warming-methane-leaks-from-all-fossil-fuels-coal-oil-and-natural-gas/>

McCurry, Justin. 'Japan fears food contamination as battle to cool nuclear plant continues'. *Guardian* [online edition], 22 March 2011 <http://www.guardian.co.uk/world/2011/mar/22/japan-food-contamination-nuclear-plant>

McIntyre, Steve. 'Briffa on another Mann hockey stick'. *Climate Audit* [online blog], 22 November 2009 <http://climateaudit.org/2009/11/22/briffa-on-another-mann-hockey-stick/>

————. 'Climategate: A battlefield perspective. Annotated notes for presentation to Heartland Conference, Chicago'. *Climate Audit* [online blog], 16 May 2010 <http://www.climateaudit.info/pdf/mcintyre-heartland_2010.pdf>

Merchant, Brian. 'Is the Gulf oil spill really *that* big of a deal?' *Treehugger* [online blog], 4 May 2010 <http://www.treehugger.com/corporate-responsibility/is-the-gulf-oil-spill-really-emthatem-big-of-a-deal.html>

'Milk spill threatens fish'. *BBC News World Edition* [online], 24 July 2002 <http://news.bbc.co.uk/2/hi/uk_news/england/2149856.stm>

Miller, Talea. 'Visualizing Japan's power outages after earthquake, tsunami'. *PBS NewsHour* [The Rundown news blog], 21 March 2011 <http://www.pbs.org/newshour/rundown/2011/03/electricity-losses-from-japan-earthquake-tsunami.html>

Milloy, Steve. 'Climategate 2.0: Jones, Briffa say Mann, hokey stick "on dodgy ground"'. *JunkScience.com* [online blog], 28 November 2011 <http://junkscience.com/2011/11/28/climategate-2-0-jones-briffa-say-mann-hokey-stick-on-dodgy-ground/>

Mitchell, R., I.R. MacDonald and K. Kvenvolden. 'Estimates of total hydrocarbon seepage into the Gulf of Mexico based on satellite remote sensing images'. *EOS Supplement* 80 (OS242), 1999

'Modified version of original post written by Josef Oehmen'. *MIT NSE Nuclear Information Hub* [online blog], 13 March 2011 <http://mitnse.com/2011/03/13/modified-version-of-original-post/>

Møller, A. P., et al. 'Dispersal and climate change: A case study of the Arctic tern *Sterna paradisaea*'. *Global Change Biology* 12 (10), October 2006,

pp. 2005–2013 <http://onlinelibrary.wiley.com/doi/10.1111/
j.1365-2486.2006.01216.x/full>

Monaghan, Sarah. 'Can oil and wildlife mix?' *Geographical* [online edition],
July 2008 <http://www.geographical.co.uk/Magazine/Gabon_-_July_08.
html>

Monbiot, George. 'Why Fukushima made me stop worrying and love
nuclear power'. *Guardian* [online edition], 21 March 2011 <http://
www.guardian.co.uk/commentisfree/2011/mar/21/pro-nuclear-japan-
fukushima>

Montford, Andrew. 'Climate cuttings 33'. *Bishop Hill* [online blog], 20
November 2009 <http://bishophill.squarespace.com/blog/2009/11/20/
climate-cuttings-33.html>

———. 'The Yamal deception'. *Bishop Hill* [online blog], 9 May 2012
<http://www.bishop-hill.net/blog/2012/5/9/the-yamal-deception.html>

Morriss, Andrew. 'Survival of the sea turtle: Cayman Turtle Farm starts
over'. *Property and Environment Research Center Reports* 24 (3), Fall 2006
<http://www.perc.org/articles/article825.php>

Morton, Oliver. 'Heroes of the environment: James Lovelock'. *TIME* Special
[*TIME* online], 17 October 2007 <http://www.time.com/time/
specials/2007/article/0,28804,1663317_1663319_1669893,00.html>

Mosher, Steven W. 'President Obama's bizarre "science czar": Dr John
R. Holdren, professional alarmist'. *PRI Review* 19 (5), September/October
2009 <http://www.pop.org/content/president-obamas-bizarre-science-
czar--dr-1958>

Mouland, Bill. 'Global warming sees polar bears stranded on melting ice'.
Mail Online [*Daily Mail* online edition], 1 February 2007 <http://www.
dailymail.co.uk/news/article-433170/Global-warming-sees-polar-bears-
stranded-melting-ice.html>

Muller, Richard A. 'Global warming bombshell: A prime piece of evidence
linking human activity to climate change turns out to be an artefact of
poor mathematics'. *MIT Technology Review* [online edition], 15 October
2004 <http://www.technologyreview.com/energy/13830/>

NASA/Goddard Space Flight Center–EOS Project Science Office.
'Scientists find that tons of oil seep into the Gulf of Mexico each year'.
ScienceDaily [online news site], 27 January 2000 <http://www.
sciencedaily.com/releases/2000/01/000127082228.htm>

Ncube, Sarah. 'Zimbabwe farmers calls for planting of GMOs'. *Zimbabwe*

Telegraph [via Africa Agriculture News], 19 November 2009 <http://
www.africa-agri.com/zimbabwe-farmers-calls-for-planting-of-gmos/>
'*New York Times* quietly edits article about Fukushima evacuation'. *Japan Probe*
[online blog], 25 March 2011 <http://www.japanprobecom/2011/03/26/
new-york-times-quietly-edits-article-about-fukushima-evacuation/>
Niedbala, Bob. 'Tributary in trouble'. *Observer-Reporter* [online edition],
6 March 2010 <http://www.observer-reporter.com/or/story11/06-03-
2010-mon-river-makes-list>
'Nigeria destroys 500 illegal oil refineries in Niger Delta'. *Platts* [online news
site], 7 March 2011 <http://www.platts.com/RSSFeedDetailedNews/
RSSFeed/Oil/8624484>
Njobeni, Siseko. 'Karoo's shale gas to boost economy, says report'. *Business
Day* [online news], 2 March 2012 <http://www.businessday.co.za/
articles/Content.aspx?id=166487>
'Nuclear radiation "the greatest public health hazard"'. *CNN International*
[online], 25 March 2011 <http://edition.cnn.com/2011/
OPINION/03/25/caldicott.nuclear.health/index.html>
Osborn, Stephen G., et al. 'Methane contamination of drinking water
accompanying gas-well drilling and hydraulic fracturing'. *Proceedings of
the National Academy of Sciences* 108 (20), 17 May 2011, pp. 8172–8176
<http://www.pnas.org/content/108/20/8172.full>
Osburn, Max R. 'Effect of DDT on the little fire ant'. *Proceedings of the Florida
State Horticultural Society*, 1945, pp. 156–158 <http://www.fshs.org/
Proceedings/Password%20Protected/1945%20Vol.%20
58/156-158(OSBURN).pdf>
Paquette, Carole. 'New species revives fears of a gypsy moth infestation'.
New York Times [online archives], 25 September 1994 <http://www.
nytimes.com/1994/09/25/nyregion/new-species-revives-fears-of-a-
gypsy-moth-infestation.html>
Paster, Pablo. 'Ask Pablo: Why didn't they just burn the BP oil spill?'
Treehugger [online blog], 3 May 2010 <http://www.treehugger.com/
natural-sciences/ask-pablo-why-didnt-they-just-burn-the-bp-oil-
spill.html>
Pearson, Sophia. 'Polar bear's listing as threatened species upheld by
U.S. federal judge'. *Bloomberg* [online news], 30 June 2011
<www.bloomberg.com/news/2011-06-30/polar-bear-s-listing-as-
threatened-upheld-in-lawsuit-over-economic-effects.html>

Pielke, Roger, Jr. 'GRL and James Saiers'. *Roger Pielke Jr's Blog* [online blog], 28 November 2009 <http://rogerpielkejr.blogspot.com/2009/11/grl-and-james-saiers.html>

'Plenty of gloom: Forecasters of scarcity and doom are not only invariably wrong, they think that being wrong proves them right'. *The Economist* [online], 18 December 1997 <http://www.economist.com/node/455855>

'Plutonium in the environment'. *MIT NSE Nuclear Information Hub* [online blog], misdated 18 March 2011 <http://mitnse.com/2011/03/18/323/>

Prasad, Bhaskar. 'Power Balance bracelet company files for bankruptcy'. *International Business Times* [online news site], 21 November 2011 <http://www.ibtimes.com/articles/253724/20111121/power-balance-bracelet-company-files-bankruptcy.htm>

Pritchard, Justin. 'BP rig inspections were fewer than advertised'. Associated Press [on *msnbc.com* online news site], 16 May 2010 <http://www.msnbc.msn.com/id/37179888/ns/disaster_in_the_gulf/t/bp-rig-inspections-were-fewer-advertised/>

Purefoy, Christian. 'Death and oil in Niger Delta's illegal refineries'. *CNN World* [online], 3 August 2010 <http://articles.cnn.com/2010-08-03/world/nigeria.oil.niger.delta_1_heating-oil-oil-industry-crude?_s=PM:WORLD>

Quiggan, John. 'Rehabilitating Carson'. *Prospect* [online edition], 24 May 2008 <http://www.prospectmagazine.co.uk/2008/05/rehabilitatingcarson/>

Rademacher, Paul. 'How big is the Deepwater Horizon oil spill?' *Google Lat Long Blog*, 7 May 2010 <google-latlong.blogspot.com/2010/05/how-big-is-deepwater-horizon-oil-spill.html>

Radul, Vern. 'Deepwater Horizon: The Gulf Stream is only the beginning'. *Antemedius* [online blog], 5 May 2010 <http://antemedius.com/content/deepwater-horizon-gulf-stream-only-beginning>

'Raising the awareness of the African public about the impact of GMOs'. Environmental Rights Action: Friends of the Earth, Nigeria [blog post] <http://www.eraction.org/component/content/24?task=view>

Rasool, S.I., and S.H. Schneider. 'Atmospheric carbon dioxide and aerosols: Effects of large increases on global climate'. *Science* 173 (3992), 9 July 1971, pp. 138–141 <http://www.sciencemag.org/content/173/3992/138.abstract>

Raymond, Eric S. 'Hiding the decline: Part 1 – The adventure begins'. *Armed*

and Dangerous [online blog], 24 November 2009 <http://esr.ibiblio.
org/?p=1447>

Reddy, Christopher. 'How science failed during the Gulf oil disaster'. Wired.
com reprinted in *ArsTechnica* [online], 22 April 2012 <http://arstechnica.
com/science/news/2012/04/how-science-failed-during-the-gulf-oil-
disaster.ars>

Regis, Ed. 'The doomslayer'. *Wired* [online edition], May 2002 <www.wired.
com/wired/archive/5.02/ffsimon_pr.html>

Reiter, Paul. 'Dangers of disinformation'. *International Herald Tribune*
Opinion, 11 January 2007 <http://www.nytimes.com/2007/01/11/
opinion/11iht-edreiter.4171294.html>

Reuters. 'Alarm over plutonium'. *Times Live* [online news site], 29 March 2011
<http://www.timeslive.co.za/world/article994217.ece/Alarm-over-
plutonium>

Revkin, Andrew C. 'A climate scientist who engages skeptics'. *New York Times*
[online Opinion Page], 27 November 2009 <http://dotearth.blogs.
nytimes.com/2009/11/27/a-climate-scientist-on-climate-skeptics/>

Richard, Michael Graham. 'Ford saves $1.2 million and reduces CO2
emissions by around 20,000 tons by turning computers off'. *Treehugger*
[online blog], 23 March 2010 <http://www.treehugger.com/gadgets/
ford-saves-12-million-and-reduces-co2-emissions-by-around-20000-
tons-by-turning-computers-off.html>

———. 'Gulf of Mexico oil spill: The what, when and where'. *Planet Green*
[online blog], 3 June 2010 <http://planetgreen.discovery.com/tech-
transport/gulfofmexico-oilspill-whatwhenwhere-whatyoucando.html>

———. 'The "Katrina of smell" is attacking News Orleans thanks to BP oil
spill'. *Treehugger* [online blog], 30 April 2010 <http://www.treehugger.
com/corporate-responsibility/the-katrina-of-smell-is-attacking-news-
orleans-thanks-to-bp-oil-spill.html>

Richards, Zoe T., et al. 'Bikini Atoll coral biodiversity resilience five decades
after nuclear testing'. *Marine Pollution Bulletin* 56 (3), March 2008, pp.
503–515 <http://www.sciencedirect.com/science/article/pii/
S0025326X07004523>

Rosenthal, Elizabeth, and Andrew C. Revkin. 'Science panel calls global
warming "unequivocal"'. *New York Times*, 3 February 2007 <http://www.
nytimes.com/2007/02/03/science/earth/03climate.html?ex=132815880
0&en=3a845c84e21df549&ei=5088&partner=rssnyt&emc=rss>

Sand, Jordan. 'A short history of MSG: Good science, bad science and taste cultures'. *Gastronomica* 5 (4), 2005, pp. 38–49 <http://www.jstor.org/pss/10.1525/gfc.2005.5.4.38>

Sands, Sarah. 'We're all doomed! 40 years from global catastrophe – and there's NOTHING we can do about it, says climate change expert'. *Mail Online* [*Daily Mail* online edition], 22 March 2008 <http://www.dailymail.co.uk/news/article-541748/Were-doomed-40-years-global-catastrophe--theres-NOTHING-says-climate-change-expert.html>

SAPA. 'Rhino poachers jailed for 25 years each'. *Times Live* [online news site], 1 February 2012 <http://www.timeslive.co.za/local/2012/02/01/rhino-poachers-jailed-for-25-years-each>

Schneider, Stephen H. 'Don't bet all environmental changes will be beneficial'. *American Physical Society News* 5 (8), August/September 1996, p. 5 <http://www.americanphysicalsociety.com/publications/apsnews/199608/upload/aug96.pdf>

Schneiderman, Jill. 'Pandora's oil well'. *EarthDharma* [online blog], 11 May 2010 <http://jillschneiderman.wordpress.com/2010/05/11/pandoras-well/>

Shackford, Stacey. 'Natural gas from fracking could be "dirtier" than coal, Cornell professors find'. *Chronicle Online* [Cornell University online news site], 11 April 2011 <http://www.news.cornell.edu/stories/April11/GasDrillingDirtier.html>

'Shell enters Karoo shale basin'. *Oil & Gas Insight* [online newsletter], December 2009 <http://www.oilandgasinsight.com/file/84725/shell-enters-karoo-shale-basin.html>

Sherwell, Philip. 'BP oil spill: Dramatic recovery of Gulf of Mexico one year on'. *The Telegraph* [online news site], 10 April 2011 <http://www.telegraph.co.uk/finance/newsbysector/energy/oilandgas/8423173/BP-oil-spill-Dramatic-recovery-of-Gulf-of-Mexico-one-year-on.html>

———. 'Louisiana oil spill may be five times bigger than previously thought'. *The Telegraph* [online news site], 1 May 2010 <http://www.telegraph.co.uk/news/worldnews/northamerica/usa/7664907/Louisiana-oil-spill-may-be-five-times-bigger-than-previously-thought.html>

Simon, Julian L. 'The statistical flummery of species loss'. 13 October 1993 <http://www.juliansimon.com/writings/Articles/SPECIFOR.txt.>

Simpson, D.W. 'Triggered earthquakes'. *Annual Review of Earth and Planetary*

Sciences 14, May 1986, pp. 21–42 <http://www.annualreviews.org/doi/abs/10.1146/annurev.ea.14.050186.000321>

Singh, Mayshree, et al. 'Seismotectonic models for South Africa: Synthesis of geoscientific information, problems, and the way forward'. *Seismological Research Letters* 80 (1), Jan/Feb 2009, pp. 71–80 <http://www.africaarray.psu.edu/publications/pdfs/Singh.pdf>

Sirico, Robert A. 'Receiving the gift of environmental stewardship'. *Acton Institute for the Study of Religion and Liberty* [online], 3 May 2000 <http://conservation.catholic.org/gift_of_stewardship.htm>

Slanina, Sjaak. 'Air pollution in China'. *The Encyclopedia of Earth* [online reference], 13 May 2008 <http://www.eoearth.org/article/Air_pollution_in_China>

'Snap! Freezing bears'. ABC Australia's *Media Watch* [online], 2 April 2007 <http://www.abc.net.au/mediawatch/transcripts/s1887890.htm>

Sonnenschein, Carlos, and Ana M. Soto. 'Don't blame the usual suspect for cancer'. *New Scientist* (2842), 10 December 2011, pp. 30–31 <http://www.newscientist.com/article/mg21228425.600-dont-blame-the-usual-suspect-for-cancer.html

Späth, Andreas. 'Much ado about fracking'. *News24* [online news site], 20 April 2011 <http://www.news24.com/Columnists/AndreasSpath/Much-ado-about-fracking-20110420>

spisharam. 'Almost 25 years later, Chernobyl and its effect on biodiversity'. *Connect-Green*, 2 August 2010 <http://www.connect-green.com/almost-25years-later-chernobyl-its-effect-on-biodiversity/>

Tabuchi, Hiroko, et al. 'Japan encourages a wider evacuation from reactor area'. *New York Times*, 26 March 2011 <http://www.nytimes.com/2011/03/26/world/asia/26japan.html>

'The 2011 Virginia earthquake'. *MIT NSE Nuclear Information Hub* [online blog], 12 September 2011 <http://mitnse.com/2011/09/12/the-2011-virginia-earthquake/>

'The WSJ guide to climate change'. *Wall Street Journal* [online edition], 10 March 2010 <http://online.wsj.com/article/SB10001424052748704007804574574101605007432.html>

'Tremor hits southern Cape'. *News24* [online news site], 14 May 2011 <http://www.news24.com/SouthAfrica/News/Tremor-hits-Southern-Cape-20110514>

Turner, Tom. 'The vindication of a public scholar'. *Earth Island Journal* 24 (2),

2009 <http://www.earthisland.org/journal/index.php/eij/article/the_
vindication_of_a_public_scholar/>

UPI. 'More biodiversity at Chernobyl'. *PhysOrg* [online news site], 12 August
2005 <http://www.physorg.com/news5774.html>

'Vaccines: The case of measles'. *Nature* (473), 23 May 2011, pp. 434–435
<http://www.nature.com/news/2011/110525/full/473434a.html>

Van der Zee, Bibi. 'Japan nuclear crisis puts UK public off new power stations'.
Guardian [online edition], 22 March 2011 <http://www.guardian.co.uk/
environment/2011/mar/22/japan-nuclear-crisis-uk-power-stations>

Vecchiatto, Paul. 'Shell could spend $200m on Karoo gas exploration'.
Business Live [online news site], 25 March 2011 <http://www.
businesslive.co.za/southafrica/sa_generalnews/2011/03/25/shell-could-
spend-200m-on-karoo-gas-exploration>

Vegter, Ivo. 'A fracture in SA's astronomical advantage?' *ITWeb Brainstorm*,
7 June 2011 <http://www.brainstormmag.co.za/index.php?option=com_
content&view=article&id=4228:a-fracture-in-sas-astronomical-
advantage>

———. 'Climate clarity'. *Daily Maverick* [online news site], 29 December
2009 <http://dailymaverick.co.za/opinionista/2009-12-29-climate-
clarity>

———. 'Climate fraud kills people'. *Daily Maverick* [online news site],
1 December 2009. http://dailymaverick.co.za/opinionista/2009-12-01-
Climate-fraud-kills-people

———. 'Drowning in calamity'. *ITWeb Brainstorm*, 1 September 2008
<http://www.brainstormmag.co.za/index.php?option=com_content&vie
w=article&id=233:drowning-in-calamity&Itemid=128>

———. 'Green tech: Doubling down on a losing bet'. *Daily Maverick* [online
news sits], 31 January 2011 <http://dailymaverick.co.za/
opinionista/2012-01-31-green-tech-doubling-down-on-a-losing-bet>

———. 'How to exploit polar bears'. *The Spike* [online blog], 19 May 2008
<http://ivo.co.za/2008/05/19/how-to-exploit-polar-bears/>

———. 'If polar bears are doomed, we all are'. *The Spike* [online blog],
6 February 2008 <http://ivo.co.za/2008/02/06/if-polar-bears-are-
doomed-we-all-are/>

———. 'If they want rhino horn, let's sell them some'. *Daily Maverick* [online
news site], 4 October 2011 <http://dailymaverick.co.za/
opinionista/2011-10-04-if-they-want-rhino-horn-lets-sell-them-some>

————. 'I'm ashamed for my profession'. *Daily Maverick* [online news site], 4 April 2011 <http://dailymaverick.co.za/opinionista/2011-04-04-im-ashamed-for-my-profession>

————. 'Karoo fracking scandal exposed!' *Daily Maverick* [online news site], 13 April 2011 <http://dailymaverick.co.za/opinionista/2011-04-13-karoo-fracking-scandal-exposed>

————. 'No logo means carte blanche'. *Daily Maverick* [online news site], 6 June 2011 <http://dailymaverick.co.za/opinionista/2011-06-06-no-logo-means-carte-blanche>

————. 'Nuclear industry wins PR award'. *The Spike* [online blog], 20 February 2008 <http://ivo.co.za/2008/02/20/nuclear-industry-wins-pr-award/>

————. 'Pop goes the hot air balloon'. *Daily Maverick* [online news site], 24 November 2009 <http://dailymaverick.co.za/opinionista/2009-11-24-Pop-goes-the-hot-air-balloon>

————. 'TEDxJohannesburg live blog'. *The Spike* [online blog], 15 November 2009 <http://ivo.co.za/2009/11/15/tedxjohannesburg-live-blog/>

————. 'Ten reasons to reject climate alarmism'. *Daily Maverick* [online news site], 2 March 2010 <http://dailymaverick.co.za/opinionista/2010-03-02-ten-reasons-to-reject-climate-alarmism>

————. 'The climate dominoes fall'. *Daily Maverick* [online news site], 16 February 2010 <http://dailymaverick.co.za/opinionista/2010-02-16-the-climate-dominoes-fall>

————. 'The great polar bear crisis'. *The Spike* [online blog], 15 May 2008 <http://ivo.co.za/2008/05/15/the-great-polar-bear-crisis/>

————. 'Who's afraid of the nuclear wolf?' *Daily Maverick* [online news site], 15 March 2011 <http://dailymaverick.co.za/opinionista/2011-03-15-whos-afraid-of-the-nuclear-wolf>

Wald, Matthew, and David Jolly. 'Dangerous levels of radioactive isotope found 25 miles from nuclear plant'. *New York Times*, 31 March 2011 <http://www.nytimes.com/2011/03/31/world/asia/31japan.html>

Wapner, Jessica. 'Did alternative medicine extend or abbreviate Steve Jobs's life?' *Scientific American* [online edition], 27 October 2011 <http://www.scientificamerican.com/article.cfm?id=alternative-medicine-extend-abbreviate-steve-jobs-life>

Wassener, Bettina. 'As affluence spreads, so does the trade in endangered species'. *International Herald Tribune Business*, 2 January 2012 <http://

www.nytimes.com/2012/01/02/business/global/as-affluence-spreads-so-does-the-trade-in-endangered-species.html>

Watson, Paul Joseph. 'Nuclear expert: Monbiot "criminally irresponsible" for downplaying Fukushima'. Alex Jones' *Infowars.com* [online blog], 28 March 2011 <http://www.infowars.com/nuclear-expert-monbiot-criminally-irresponsible-for-downplaying-fukushima/>

Watts, Anthony. 'Climategate 2 email – Rob Wilson replicates McIntyre and McKitrick – produces hockey sticks out of noise'. *Watts Up With That?* [online blog], 27 November 2011 <http://wattsupwiththat.com/2011/11/27/climategate-2-email-briffa-replicates-mcintyre-mckitrick-produces-hockey-sticks-out-of-noise/>

———. 'Zorita calls for barring Phil Jones, Michael Mann, and Stefan Rahmstorf from further IPCC participation'. *Watts Up With That?* [online blog], 27 November 2009 <http://wattsupwiththat.com/2009/11/27/zorita-calls-for-barring-phil-jones-michael-mann-and-stefan-rahmstorf-from-further-ipcc-participation/>

Webster, Ben. 'Climate chief was told of false glacier claims before Copenhagen'. *The Times* [online edition], 30 January 2010 <http://www.timesonline.co.uk/tol/news/environment/article7009081.ece>

Whitty, Julia. 'Why is population control such a radioactive topic?' *Mother Jones* [online forum], 12 May 2010 <http://www.motherjones.com/blue-marble/2010/05/population-forum>

Wingo, P.A., et al. 'Long-term trends in cancer mortality in the United States, 1930–1998'. *Cancer* 97 (12), 15 June 2003, pp 3133–3275 <http://www.ncbi.nlm.nih.gov/pubmed/12784323>

Wood, Shauna. 'What is in a vaccine???' *Vaccination News* [online news site], 9 August 2000 <http://www.vaccinationnews.com/dailynews/may2001/whatsinvax.htm>

Yeld, John. 'R10m war chest for anti-frackers'. *Cape Argus*, 23 May 2011 <www.iol.co.za/capeargus/r10m-war-chest-for-anti-frackers-1.1072507>

Zadnik, Karla, et al. 'Vision: Myopia and ambient night-time lighting'. *Nature* 404, 9 March 2000, pp. 143–144 <http://www.nature.com/nature/journal/v404/n6774/abs/404143a0.html>

Zeratsky, Katherine. 'Monosodium glutamate (MSG): Is it harmful?' Mayo Clinic Expert Answers [online post] <http://www.mayoclinic.com/health/monosodium-glutamate/an01251>

Zvomuya, Fidelis. 'Raw and exposed: Processing'. *Dairy Mail* 17 (6), June

2010, pp. 92–93, 95 <http://www.sabinet.co.za/abstracts/ac_dm/ac_dm_
v17_n6_a21.html>

FILMS AND VIDEOS

Durkin, Martin (prod. and dir.). *The Great Global Warming Swindle.*
Documentary film, 2007 <http://www.youtube.com/
watch?v=YtevF4B4RtQ>

Fox, Josh (dir.). *Gasland.* Documentary film, 2010 <http://www.
gaslandthemovie.com/about-the-film/>

McClellan, Max (prod.). 'Can hunting endangered animals save the species?'
Segment on CBS News *60 Minutes*, 29 January 2012 <http://www.
cbsnews.com/8301-18560_162-57368000/can-hunting-endangered-
animals-save-the-species/>

Minchin, Tim (writer) and Tracy King (prod.). *Storm.* Animated short film,
2011 <http://www.timminchin.com/2011/04/08/storm/>

'Penn & Teller get hippies to sign water banning petition'. Video uploaded
on YouTube by supersteve9219, 6 December 2006 <http://www.youtube.
com/watch?v=yi3erdgVVTw>

'Radiation risk from nuclear power station in Japan'. Segment on *BBC News*,
14 March 2011 <http://www.dailymotion.com/video/xhlgks_radiation-
risk-from-nuclear-power-station-in-japan_news>

Shaw, Neil (prod.). 'Nuke'. Segment on M-Net's *Carte Blanche*, 27 June 2010
<http://beta.mnet.co.za/carteblanche/Article.
aspx?Id=4012&ShowId=1>

LECTURE NOTES

Datta, Saugatta. 'Groundwater Hydrology'. Lecture notes for Hydrogeology
GEOL 611, Kansas State University, 2011 <http://hercules.gcsu.
edu/~sdatta/home/teaching/hydro/lectures/
Groundwater%2520Hydrology.ppt>

Möller, Gregory. 'The environmental history of DDT and *Silent Spring*'.
Lecture notes for Principles of Environmental Toxicology, University of
Idaho <http://www.agls.uidaho.edu/etox/lectures/lecture02/Slides_
SILENTSPRING.pdf>

Skone, Timothy J. 'National Energy Technology Laboratory: Life cycle
greenhouse gas analysis of natural gas extraction and delivery in the
United States'. Cornell University lecture series, 12 May 2011

<http://cce.cornell.edu/EnergyClimateChange/NaturalGasDev/
Documents/PDFs/SKONE_NG_LC_GHG_Profile_Cornell_12MAY11_
Final.pdf>

Stute, Martin. 'Darcy's Law'. Lecture notes for Hydrology EESC BC ENV
3025, Columbia University, 2011 <http://www.ldeo.columbia.
edu/~martins/hydro/lectures/darcy.html>

PRESS RELEASES AND STATEMENTS

Adatia, Rachel, and Earthlife Africa Cape Town. 'Anti-fracking activists call
for a national boycott of Shell!' 16 May 2011 <http://www.earthlife.org.
za/?p=1591>

Chevron. 'Chevron opens largest wastewater treatment plant in SA'. 22
August 2008 <http://www.petroleumafrica.com/en/newsarticle.
php?NewsID=6345>

Crowell & Moring. 'Update on the polar bear–Endangered Species Act
multidistrict litigation'. 31 October 2011 <http://www.crowell.com/
NewsEvents/AlertsNewsletters/Environment-Natural-Resources-Law-
Alert/Update-on-the-Polar-Bear-Endangered-Species-Act-
Multidistrict-Litigation>

Defenders of Wildlife. 'Polar bear litigation'. 12 November 2008 <http://
www.defenders.org/polar-bear-litigation>

Econometrix. 'Economic report: Karoo shale gas development could boost
GDP and create hundreds of thousands of jobs'. 2 March 2012 <http://
econometrix.co.za/DealerNews/newsdetails.asp?newsitem=6534>

Eisenhower Foundation. 'Eisenhower and the righteous cause: The liberation
of Europe'. 2010 <http://www.dwightdeisenhower.com/pdfs/righteous_
cause_prospectus.pdf>

Fox, Josh. 'Affirming *Gasland*: A de-debunking document in response to
specious and misleading gas industry claims against the film'. July 2010
<http://lockthegate.org.au/documents/doc-237-affirming-gasland.pdf>

Greenpeace Africa. 'The Advanced Energy [R]evolution'. 24 May 2011
<http://www.greenpeace.org/africa/en/News/news/The-Advanced-
Energy-Revolution-Report/>

Greenpeace International. 'South Africa: Fukushima accident proves that
nuclear is dangerous – Greenpeace'. 27 February 2012 <http://allafrica.
com/stories/201202270798.html>

Howarth, Robert W. 'Assessment of the greenhouse gas footprint of natural gas from shale formations obtained by high-volume, slick-water hydraulic fracturing'. 11 April 2011 <http://www.eeb.cornell.edu/howarth/GHG%20update%20--%20April%2011%202011.pdf>

Jemmi, Thomas. 'CITES a success story: The Swiss view'. Speech on the thirty-fifth anniversary of CITES, 1 July 2010 <http://www.cites.org/common/news/2010/E-CITES_35_birthday-OFV.pdf>

National Chamber Litigation Center. 'In re: Polar bear Endangered Species Act listing and 4(d) litigation'. 16 August 2010 <http://www.chamberlitigation.com/re-polar-bear-endangered-species-act-listing-and-4d-litigation>

State of Colorado Oil and Gas Conservation Commission. 'Department of Natural Resources press release: *Gasland*'. 29 October 2010 <http://cogcc.state.co.us/library/GASLAND%20DOC.pdf>

University Corporation for Atmospheric Research. 'Ocean currents likely to carry oil along Atlantic coast'. 3 June 2010 <http://www2.ucar.edu/news/ocean-currents-likely-to-carry-oil-spill-along-atlantic-coast>

World Health Organization. 'WHO recommended insecticides for indoor residual spraying against malaria vectors'. October 2007 <http://www.who.int/whopes/Insecticides_IRS_Malaria_ok.pdf>

REPORTS

American Cancer Society. *Cancer Facts and Figures 2011*. Atlanta: American Cancer Society, 2001 <http://www.cancer.org/acs/groups/content/@epidemiologysurveilance/documents/document/acspc-029771.pdf>

AON Benfield. 'South Africa Spotlight on Earthquake'. June 2010 <http://www.aon.com/attachments/reinsurance/201006_mega_eq_report.pdf>

Bureau of Market Research, University of South Africa. 'Household income and expenditure patterns and trends, 2008–2009'. 24 November 2010 <www.unisa.ac.za/contents/faculties/ems/docs/Press395.pdf>

Busby, Christopher. 'Predicting the Global Health Consequences of the Chernobyl Accident: Methodology of the European Committee on Radiation Risk'. Green Audit, 24 April 2011 <http://www.euradcom.org/2011/chernhealthrept3.pdf>

Center for Advanced Nuclear Energy Systems. 'Technical lessons learned from the Fukushima-Daichii accident and possible corrective actions for the nuclear industry: An initial evaluation'. 26 July 2011 <http://mitnse.

files.wordpress.com/2011/08/fukushima-lessons-learned-mit-nsp-025_
rev1.pdf>

Committee Examining Radiation Risks of Internal Emitters. 'Report of the
Committee Examining Radiation Risks of Internal Emitters (CERRIE)'.
October 2004 <http://www.cerrie.org/pdfs/cerrie_report_e-book.pdf>

European Renewable Energy Council and Greenpeace. 'The Advanced
Energy [R]evolution: A Sustainable Energy Outlook for South Africa'.
May 2011 <http://www.energyblueprint.info/fileadmin/media/
documents/national/2011/E_R__South_Africa_May_2011-LR.pdf>

Geo Pollution Technologies, for D Light Attorneys. 'Critical Review of
Report: Evaluation of the Groundwater Report in Support of the EMP
for the South Western Karoo Basin Gas Exploration Application Project
– Eastern Precinct'. 30 March 2011 <http://www.greenkaroo.co.za/
uploads/media/Critical_evaluation_of_the_groundwater_report_in_
support_of_the_EMP_for_the_South_Western_Karoo.pdf>

Golder Associates. 'Proposed south western Karoo Basin gas exploration
project by Shell Exploration Company B.V.' 2011 <http://www.golder.
com/af/en/modules.php?name=Pages&sp_id=1236>

Greenpeace. 'The Chernobyl Catastrophe: Consequences on Human
Health'. April 2006 <http://www.greenpeace.org/international/Global/
international/planet-2/report/2006/4/chernobylhealthreport.pdf>

Haddad, Robert, and Steven Murawski. 'Analysis of Hydrocarbons in Samples
Provided from the Cruise of the R/V Weatherbird II, May 23–26, 2010'.
National Oceanic and Atmospheric Administration, 2010 <http://www.
noaanews.noaa.gov/stories2010/PDFs/noaa_weatherbird_analysis.pdf>

Haken, Jeremy. 'Transnational Crime in the Developing World: A February
2011 Report from Global Financial Integrity,' February 2011 <http://
transcrime.gfintegrity.org/>

Havemann Inc. Specialist Energy Attorneys, for Treasure the Karoo Action
Group. 'A Critical Review of the Application for a Karoo Gas
Exploration Right by Shell Exploration Company B.V.' 5 April 2011
<http://havemanninc.com/wp-content/uploads/2011/04/
KarooPolicyObjection-5-April-0903-FINAL.pdf>

International Atomic Energy Agency. 'IAEA International Fact Finding
Expert Mission of the Nuclear Accident Following the Great East Japan
Earthquake and Tsunami'. 1 June 2011 <http://www.iaea.org/
newscenter/focus/fukushima/missionsummary010611.pdf>

International Energy Agency. 'World Energy Outlook 2011: Special Report: Are We Entering a Golden Age of Gas?' 2011 <http://www.iea.org/weo/docs/weo2011/WEO2011_GoldenAgeofGasReport.pdf>

McKitrick, Ross. 'What is the "hockey stick" debate about?' University of Guelph, 4 April 2005 <http://www.uoguelph.ca/~rmckitri/research/McKitrick-hockeystick.pdf>

Milliken, Tom, et al. 'African and Asian Rhinoceroses – Status, Conservation and Trade: A Report from the IUCN Species Survival Commission (IUCN/SSC) African and Asian Rhino Specialist Groups and TRAFFIC to the CITES Secretariat Pursuant to Resolution Conf. 9.14 (Rev. CoP14) and Decision 14.89'. 20 November 2009 <http://www.cites.org/common/cop/15/doc/E15-45-01A.pdf>

Monfils, Rean. 'The Global Risk of Marine Pollution from WWII Shipwrecks: Examples from the Seven Seas'. Sea Australia <http://www.seaaustralia.com/documents/The%20Global%20Risk%20of%20Marine%20Pollution%20from%20WWII%20Shipwrecks-final.pdf>

Morano, Marc. 'Climate Depot Special Report: More Than 1000 International Scientists Dissent Over Man-Made Global Warming Claims – Challenge UN IPCC and Gore'. 8 December 2010 <http://www.climatedepot.com/a/9035/SPECIAL-REPORT-More-Than-1000-International-Scientists-Dissent-Over-ManMade-Global-Warming-Claims--Challenge-UN-IPCC--Gore>

National Commission on the BP Deepwater Horizon Oil Spill and Offshore Drilling. 'Final Report to the President: Deepwater: The Gulf Oil Disaster and the Future of Offshore Drilling'. January 2011 <www.oilspillcommission.gov/sites/default/files/documents/DEEPWATER_ReporttothePresident_FINAL.pdf>

Operational Science Advisory Team (OSAT-2) for US Coast Guard. 'Summary Report for Fate and Effects of Remnant Oil Remaining in the Beach Environment'. 10 February 2011 <http://www.restorethegulf.gov/sites/default/files/u316/OSAT-2%20Report%20no%20ltr.pdf>

Sas-Rolfes, Michael 't. 'Saving African rhinos: A market success story'. *Property and Environment Research Center Case Studies Series*, August 2011 <http://www.perc.org/files/Saving%20African%20Rhinos%20final.pdf>

UK Environment Alliance. 'Pollution Prevention Guidelines: Dairies and Other Milk Handling Operations PPG17'. 2001 <www.doeni.gov.uk/niea/ppg17.pdf>

UK House of Commons Energy and Climate Change Committee. 'Shale Gas: Fifth Report of Session 2010–12'. 23 May 2011 <http://www.publications.parliament.uk/pa/cm201012/cmselect/cmenergy/795/795.pdf>

UN Chernobyl Forum Expert Group 'Health'. 'Health Effects of the Chernobyl Accident and Special Health Care Programmes'. 2006 <http://whqlibdoc.who.int/publications/2006/9241594179_eng.pdf>

UN Department of Economic and Social Affairs/Population Division. 'World Population to 2300'. 2004 <http://www.un.org/esa/population/publications/longrange2/WorldPop2300final.pdf>

UN Environment Programme. 'Environmental Assessment of Ogoniland Report'. August 2011 <http://www.unep.org/nigeria/>

UN Food and Agriculture Organization and World Health Organization. Fourteenth report of the Joint FAO/WHO Expert Committee on Food Additives: 'Toxicological evaluation of some extraction solvents and certain other substances'. 1970 <http://www.inchem.org/documents/jecfa/jecmono/v48aje09.htm>

UN Global Seismic Hazard Assessment Program. 'Global seismic hazard map' assembled by D. Giardini, et al., 1999 <http://mitnse.files.wordpress.com/2011/09/globalseismichazardmap.jpg>

UN Office for the Coordination of Humanitarian Affairs. 'Japan: Earthquake & Tsunami, Situation Report No. 10'. 21 March 2011 <http://reliefweb.int/sites/reliefweb.int/files/resources/01AC3A641A6AD092C125785A0036840D-Full_Report.pdf>

UN Scientific Committee on the Effects of Atomic Radiation. 'The Chernobyl Accident: UNSCEAR's Assessments of the Radiation Effects'. 2008 <http://www.unscear.org/unscear/en/chernobyl.html>

US Department of Agriculture Forest Service. 'Slow the Spread: A National Program to Manage the Gypsy Moth'. April 2007 <http://nrs.fs.fed.us/pubs/gtr/gtr_nrs6.pdf>

US Department of Energy, Office of Fossil Energy, National Energy Technology Laboratory. 'Modern Shale Gas Development in the United States: A Primer'. April 2009 <http://fossil.energy.gov/programs/oilgas/publications/naturalgas_general/Shale_Gas_Primer_2009.pdf>

US Energy Information Administration. 'World Shale Gas Resources: An Initial Assessment of 14 Regions Outside the United States'. April 2011 <http://www.eia.gov/analysis/studies/worldshalegas/pdf/fullreport.pdf>

UN Environment Programme. 'Updated Scientific Report on the Environmental Effects of the Conflict between Iraq and Kuwait'. 10–21 May 2003 <www.unep.org/dewa/westasia/data/Knowledge_Bases/Iraq/Reports/UNEPGCIraq1993.pdf>

US Fish and Wildlife Service, et al. 'Deepwater Horizon Response Consolidated Fish and Wildlife Collection Report'. 2 November 2010 <http://www.restorethegulf.gov/sites/default/files/documents/pdf/Consolidated%20Wildlife%20Table%20110210.pdf>

Waller, Roger M. 'Groundwater and the Rural Homeowner'. US Geological Survey, 1994 <http://pubs.usgs.gov/gip/gw_ruralhomeowner/pdf/gw_ruralhomeowner.pdf>

WEBSITES

Agency for Toxic Substances and Disease Registry. 'ToxFAQs™ for 2-butoxyethanol and 2-butoxyethanol acetate' <http://www.atsdr.cdc.gov/toxfaqs/tf.asp?id=346&tid=61>

AJINOMOTO®. 'The birth of AJI-NO-MOTO®' <http://www.ajinomoto.com/features/aji-no-moto/en/basic/index.html>

An Englishman's Castle. 'READ ME for Harry's work on the CRU TS2.1/3.0 datasets, 2006-2009!' <http://www.anenglishmanscastle.com/HARRY_READ_ME.txt>

Aquatic Community. 'Yacare caiman' <http://www.aquaticcommunity.com/caimans/yacare.php>

Brian Deer Selected Investigations and Journalism. 'The Lancet scandal' <http://briandeer.com/mmr-lancet.htm>

Chase SHELL OIL out of the Karoo! Facebook Group: The official organised resistance to FRACKING in South Africa started in January 2011 by Jonathan Deal of TKAG <https://www.facebook.com/groups/chaseshelloutofthekaroo/>

Chemicals-technology.com. 'Gas to liquids (GTL), Australia' <http://www.chemicals-technology.com/projects/gtl/>

Christopher Busby Foundation for Children of Fukushima <http://www.cbfcf.org/>

City of Cape Town. 'Milk quality & safety' <http://www.capetown.gov.za/EN/CITYHEALTH/ENVIROHEALTH/FOODQUALITYANDSAFETY/Pages/MilkQualitySafety.aspx>

Climategate 2 FOIA 2011 Searchable Database <http://foia2011.org/>

Dangers of Fracking <http://www.dangersoffracking.com/>

Environmentalists for Nuclear Energy™. 'What is Gaia? by James Lovelock'
  <http://www.ecolo.org/lovelock/what_is_Gaia.html>

Exotic Wildlife Association. 'Death warrant on exotic species' <http://
  myewa.org/Original_Backup/www/forms/Death%20Warrent%20
  on%20Exotic%20Species.pdf>

Fallacy files <http://www.fallacyfiles.org/>

Federation of Electric Power Companies of Japan. 'Why is nuclear energy
  necessary in Japan?' <http://www.fepc.or.jp/english/nuclear/energy_
  policy/necessary/index.html>

Food-Info. 'Monosodiumglutamate – E621' <http://www.food-info.net/uk/
  intol/msg.htm>

FracFocus Chemical Disclosure Registry <http://fracfocus.org/>

Generation Rescue. 'Vaccine ingredients and side effects' <http://www.
  generationrescue.org/resources/vaccination/vaccine-ingredients-and-
  side-effects/>

Geology.com. 'Marcellus Shale – Appalachian Basin natural gas play'
  <http://geology.com/articles/marcellus-shale.shtml>

Groundwater Resources and Investment. 'Golf course water supply' <http://
  www.groundwatercapture.com/golfcourse/>

Gypsy Moth: An Informational Guide <http://www.gypsy-moth.com/>

International School of Well Drilling online courses <http://
  welldrillingschool.com/courses/indexNorm.aspx>

IUCN Red List of Threatened Species. 'Diceros bicornis' <http://www.
  iucnredlist.org/apps/redlist/details/6557/0>

Lewis Pugh <http://www.lewispugh.com>

Mating Habits of Arctic Terns <http://matinghabitsofarcticterns.blogspot.
  com/>

Murray State University. 'Luddites and Luddism history' <http://campus.
  murraystate.edu/academic/faculty/kevin.binfield/luddites/
  LudditeHistory.htm>

National Network for Immunization Information. 'History and achievements'
  <http://www.immunizationinfo.org/parents/why-immunize/history-
  and-achievements>

NOAA National Geophysical Data Center. 'NOAA/WDC tsunami event

database' <http://www.ngdc.noaa.gov/nndc/struts/
form?t=101650&s=70&d=7>

Online Ethics Center for Engineering and Research' <http://www.
onlineethics.org/>

Power Balance <http://www.powerbalance.com>

Rhino Economics <http://www.rhino-economics.com/>

Rhino Resource Center <http://www.rhinoresourcecenter.com/>

Save Bantamsklip <http://www.savebantamsklip.org/>

South Africa Golf Clubs <http://www.south-africa-golf-clubs.com/>

South African Government Information. 'The New Growth Path' <http://
www.info.gov.za/aboutgovt/programmes/new-growth-path/index.html>

TED.com. 'Lewis Pugh: Coldwater swimmer' <http://www.ted.com/
speakers/lewis_pugh.html>

Texas Center for Policy Studies. 'Texas environmental almanac online – 1995'.
<http://www.texascenter.org/almanac/TXENVALMANAC.HTML>

The Nizkor Project <http://www.nizkor.org/>

Tigers in Crisis. 'Traditional Chinese medicine and tigers' <http://www.
tigersincrisis.com/traditional_medicine.htm>

Treasure the Karoo Action Group <http://treasurethekaroo.co.za/>

Truth In Labeling. 'History of invention and use of MSG' <http://www.
truthinlabeling.org/IVhistoryOfUse.html>

UN Environment Programme Environmental Data Explorer <http://
geodata.grid.unep.ch/>

World Health Organization. 'Cancer fact sheet no. 297, October 2011'
<http://www.who.int/mediacentre/factsheets/fs297/en/>

World Life Expectancy. 'The History of Life Expectancy' <http://www.
worldlifeexpectancy.com/history-of-life-expectancy>

World Nuclear Association. 'Hiroshima, Nagasaki, and subsequent weapons
testing' <http://www.world-nuclear.org/info/inf52.html>

———. 'Naturally-occurring radioactive materials (NORM)' <http://
world-nuclear.org/info/inf30.html>

Wyoming Game and Fish Department. 'Hunting regulations' <http://wgfd.
wyo.gov/web2011/HUNTING-1000179.aspx>

INDEX

acid rain 217–218
Ahearn, Bill 93–94, 95
Ajinomoto Company 98, 101
Alexander, William 209
Algeria 208
Alpha well blowout 154
An Inconvenient Truth 89
antelope 250
Arctic climate 15, 256, 258
argument
   ad hominem attacks 166–172
   appeal to authority 182–183
   appeal to consensus 183–185
   appeal to motive 185–186
   art of 162–165
   denialism 164, 189–191
   and environmentalism 161–162,
      163–165, 218–220
   logical fallacies 172–182
   mistaken causation 186–187
aspartame 100–101
autism 91–95

Bantamsklip nuclear power station
   119, 120

Barnett Shale 11, 63
Bennett, Geraldine 39
Bikini Atoll 127–128
Biowatch South Africa 124
Bloom, Emily 38
Bolin, David E. 33
Booker, Christopher 207
BP 15, 63, 138–139, 141, 155, 158
Bradley, Raymond 195
Brahic, Catherine 127
Briffa, Keith 202, 204
Bundu Gas and Oil 11
Burnett, Sterling 257
Busby, Christopher 120–122, 123, 129,
   130, 131, 137
Byrd, Amanda 175, 176–177

caiman 250
Calder, Nigel 193–194, 236
Caldicott, Helen 131, 133
cancer
   as cause of death 102–106
   perceived consequence of pesticides
      81, 86, 86–87
   as result of radiation 130–131, 134

Carnegie Mellon University study 60

Carrillo, Victor G. 34

Carson, Rachel 82–83, 84, 85, 225, 229

Carte Blanche 119–120, 123–125

Cato Institute 239

Cayman Turtle Farm 249–250

CFCs 217

Checks and Balances Project 28

chemicals
   DDT 83–89, 199
   MSG 96, 97–106
   use in fracking 28, 32, 35–37, 40, 46,
     47, 177–179

Chernobyl
   biodiversity controversy 128
   cancer deaths 130–131
   meltdown 113, 121, 122, 130

Chevron 47

China 10, 45, *68*, 80, 117, *132*, 263

Chivell, Wilfred 120

Christoff, Peter 190, 190–191

CITES 245–246, 249–250

Clark, Nikki 119, 120

Clean Air Task Force 60

climate change
   Arctic climate and polar bears
     256–258
   climate-change alarmism 210–212
   errors in IPCC reports 206–209, 216
   from Pugh's perspective 11, 15–16, 19
   greenhouse gases *56*, 57, 59, 60, 181,
     199, 210, 214, 258
   'hockey stick' controversy 195–200,
     201–206
   influence of methane 57
   leaked emails from University of East
     Anglia's CRU 200–206
   logical approach to issue 213–216
   misrepresentations by media 174–177,
     181–182, 193–195

political motives 236–237
   role in spread of malaria 89

coal industry
   deaths *132*, 132, 263
   pollution 2, 37–38, 55–56, *56*, 58
   radioactivity *51*, 128–129

Colorado Independent 35

Colorado Oil & Gas Conservation
   Commission (COGCC) 30–31

Convention on International Trade in
   Endangered Species *see* CITES

coral reefs 127–128, 150, 151

Cornell University study 57–61

Cousteau, Philippe, Jr 148

Cowling, Richard 119

Creamer, Martin 9

Crichton, Michael 223, 224–226, 227

Crosbie, Dan 175, 176

Cuadrilla Resources 74

culling 253–255

Daily Mail 175, 212

Daily Maverick 12, 33, 117

DDT controversy 83–89, 199

Deal, Jonathan 21, 35, 37, 39–40, 55, 61,
   62, 65, 78

Deepwater Horizon oil spill
   actual impact of spill 145–148, 155
   comparison to other oil spills
     150–153
   consequences of exaggeration 155,
     156, 158
   exaggerated media coverage 139–140,
     141–144, 147–149, 174
   well blowout 42, 138–139, 140

Deer, Brian 93, 94

Delingpole, James 200, 237, 238

denialism 164, 189–191

Der Spiegel 208

De Wit, Maarten J. 24, 53

Dolin, Viktor 128
Duke University study 38–39, 40–41,
    57, 173
Dyer Island Conservation Trust 120

Earth Day 154, 161, 236
Earth in the Balance 230
Earthlife Africa 17, 39, 171
earthquake threat
    see also Fukushima
    Koeberg and Milnerton Fault 125–126
    in South Africa 72–75
economics
    anti-capitalist and socialist sentiments
        236–240
    economic impact of environmental
        regulations 2–3, 4–5, 7–8, 259–266
    need for economic sustainability
        240–244
    potential benefits of fracking 75–78,
        179–181, 264–266
Economist 135, 219–220
Ehrlich, Paul 102–103, 225, 229,
    230–231, 233
Eisenhower, Dwight D. 189–190
Eksteen, Hennie 128
elephants 253–255
Ellis, Richard 247
Ellsworth, Aimee 31, 32
EnCana Corporation of Canada 37, 63
endangered species
    at Chernobyl 128
    CITES 245–246
    commercial breeding of antelope
        250–252
    culling of elephants 253–255
    farm-raised turtles 249–250
    polar bears 255–259
    predictions of environmentalists
        229–230

rhinoceros 246–248, 249, 254–255
    tigers 248
Endangered Species Act (ESA) 255, 258
energy crisis (SA) 136, 262–263, 264
Energy in Depth 22
Environmental Defense Center 154
environmentalism
    anti-capitalist and socialist sentiments
        236–240
    economic harm of exaggeration 3–5, 6,
        259–266
    exaggerated argument 161–162,
        164–165, 165–167, 172–182,
        218–220
    extreme positions and predictions
        221–223, 229–235, 239–240
    need for sensible stewardship 240–244
    need to be separate from state
        power 227
    as religion 225–226, 240
    resource scarcity predictions
        231–232
Environmental Protection Agency
    see EPA
environmental regulations 2–3, 4–5, 7–8,
    44–47
Environment Canada 176
EPA 36, 58, 147, 154, 218
Eskom 76–77
Exotic Wildlife Association 250–251, 252
extinction predictions 229–230
    see also endangered species
Exxon Valdez oil spill 153

Fakir, Saliem 120
Falcon Oil & Gas 11
Fells, Ian 129
Feral, Priscilla 222, 251–252
Fesmire, Mark 33
Fig, David 124

fire ants 84
fishing rights 218
Fox, Josh 4, 21, 30, 32, 35, 37, 37–38, 172, 178–179
fracking
  amount of water needed 53–55
  earthquake argument 72–75
  explanation of process 23–29
  fallacies 172–173, 177–179
  government moratorium 78–79, 263
  job creation 75–76, 77–78, 265
  misconceptions re contamination of
    groundwater 14, 23, 24–26, 27, 30–39, 40–41, 55–62
  potential benefits for South Africa 75–78, 264–266
  shale-gas resource in Karoo 9–10, 264
  shale-gas resources in world 68
  in US 9, 10, 11, 12, 32–34, 35, 38, 49, 63, 69, 75
Friends of Animals 251, 252
Fukushima
  deaths 108, 110, 131–132
  earthquake and tsunami 108–111, 126
  exaggeration of nuclear threat 5–6, 69, 116–118, 122–123, 132, 134, 135–136, 263
  tsunami effect on nuclear reactors 111–116

Gabon 155–156, 158
Gaia hypothesis 135, 225–226, 232, 233, 234
Gashchak, Sergii 128
Gasland 4, 21, 30, 31, 35, 38, 172, 177, 178
Generation Rescue 92
Geophysical Research Letters (GRL) 205
global warming
  see also climate change
  claim to consensus 184–185

denialism 190–191
'hockey stick' controversy 195–200, 201–206
misrepresentations 174–177, 181–182
Goddard Institute for Space Science (NASA) 216, 261
Goddard, Robert 187–188
Godwin's law 190
Gore, Al 89, 175, 176, 212, 225, 230
Gorski, David 92
greenhouse gas emissions 56, 57, 59, 60, 181, 199, 210, 214, 258
  see also climate change; methane
Greenpeace 6, 17, 131, 157, 179, 180, 181, 237, 239, 257, 263
Groundwater Protection Council 53, 54
Grunwald, Michael 149
Guardian 119, 120, 126, 135, 138, 207, 212
gypsy moth 84, 87–88

Hanger, John 38, 61
Hanke, Steve 239
Hansen, James 216, 225, 261
Hardin, Garrett 242
Hartnady, Chris 74
Hassett, Kevin 258
Hawking, Stephen 164, 183
Hayward, Tony 141
Hide, Miyake 97
Himalayan glaciers 206–207
Hinkley nuclear power station 119, 120
'hockey stick' controversy 195–200, 201–206
Holdren, John 233
Holocaust 189, 190, 191
Houston Chronicle 148
Howarth, Robert 58–60, 61
Hughes, Malcolm 195
Hulme, Mike 209
hunting industry 167, 222, 251–255

Hvistendal, Mara 128–129
hydraulic fracturing *see* fracking

IAEA 118, 130
Ikeda, Kikunae 97–98, 100
immunisation *see* vaccines
Ingraffea, Tony 58, 59–60, 61
Intergovernmental Panel on Climate
    Change *see* IPCC
International Atomic Energy Agency
    see IAEA
International Union for the Conservation
    of Nature *see* IUCN
IPCC
    errors in reports 205, 206–209, 216
    greenhouse gas impact 59
    influence of 'hockey stick' chart 197,
        198, 199, 202
    leaked email documents 200, 201
    Nobel Peace Prize 212
    origin 236
    predictions on sea levels 261
IUCN 230, 246, 250, 255

Jantjies, Vuyisa 79
Japan *see* Fukushima; MSG
Jaramillo, Paula 60
Jemmi, Thomas 246
Jenner, Edward 90
job creation 75–78, 179–181, 265–266
Jobs, Steve 168
Jonah gas field 62–63
Jones, Phil 204

Karoo Basin
    earthquakes 73–74
    potential of resource 67–68, 70–72
    resource of shale gas 9–10, 11, 24
    SKA project 63–67
    surface impact of fracking 62–63, 66–67

water aquifers 14, 23, 24–26
    water needs of fracking 54, 55
Karoo Shale Gas Community Forum
    79, 266
King III report 4, 260–262
King, Bill 148
King, Mervyn 4, 260–262
Klein, Naomi 138, 139, 140, 159
Koeberg nuclear power station 119, 124,
    125–126
Kwok, Robert 99
Kyoto Protocol 199

Lancet 93
Larsen, Annika 119
Lawson, Nigel 236
Lee, Joseph J., Jr 33
Legates, David 256–257
Levi, Michael 59
Limbaugh, Rush 149
Lloyd, Philip 77
Logan, Lara 251–252
Lomborg, Bjørn 229
Lovelock, James 135, 211–213,
    226, 232

malaria 82, 83, 88–89
Malthus, Thomas 225, 231
Mann, Michael 195, 197, 198, 199, 200,
    201, 202, 203, 204, 208
Marcellus Shale 12, 38, 63, 69
Markham, Mike 31, 32
McAleer, Phelim 32
McCabe, David 60, 60–61
McClure, Renee 31, 32
McIntyre, Steve 197, 198, 201, 204,
    205, 206
McKitrick, Ross 197–198, 201, 205, 206
Meacher, Michael 121
Measles *see* MMR vaccine

media
  consequences of exaggerated coverage
    155–158
  exaggerated coverage of Deepwater
    Horizon oil spill 139–140, 141–149
  exaggeration of nuclear threat at
    Fukushima 116–118, 122–123, 132,
    134, 135–136, 263
methane 30–32, 39, 40–41, 57, 58,
  59–60
Mexico *see* Deepwater Horizon oil spill
Minchin, Tim 168
mineral rights (SA) 9, 78–79, 264
Mining Weekly 9
Mironova, Natalia 130, 131
Mitchell, Roger 153
MMR vaccine 91–95
Mohale, Bonang 55
Monbiot, George 135–136, 137
Moneyweb 39
monosodium glutamate *see* MSG
Montford, Andrew 202
Moore, Patrick 237, 239
Morocco 208
mosquitoes 82, 84, 85, 86, 89
Mossgas project 24, 69, 264
MSG
  bad reputation 96
  benefits 100–101
  description 97–99
  exaggeration of reactions 99–100
Muller, Richard 198, 205
mumps *see* MMR vaccine
Myers, Norman 229, 230

NASA 211, 216, 261
National Aeronautics and Space
  Administration *see* NASA
National Center for Atmospheric
  Research *see* NCAR

National Center for Policy Analysis
  see NCPA
National Geographic 142, 145
National Network for Immunization
  Information *see* NNII
National Ocean and Atmospheric
  Administration 148
National Post 175, 176
National Resources Defense Council 257
natural gas
  see also fracking; methane
  misconceptions re pollution 55–62
  potential extent of Karoo resource
    67–68, 70–72
Nature 95, 102, 186
NCAR 142, 145
NCPA 256, 257–258
'New Growth Path' policy (SA) 179–180
News24 39
New Scientist 127, 193, 236
New York Times 61, 105, 117, 134, 175,
  187–188
Next Big Future blog 132, *132–133*
Nigeria 15, 20, 42–43
Nixon, Richard 154
NNII 90, 91
North Anna nuclear plant 126
Nuclear Industry Association of South
  Africa 125
nuclear power
  see also Fukushima
  death rates 129–132
  effects of nuclear versus coal 128–129
  exaggeration of nuclear threat 5–6,
    118–127, 133–134, 263
  reports of increase in biodiversity
    127–128
  safest form of energy 132, *132–133*
  as solution to energy crisis 262–263
  support 135–137, 236

Obama, Barack 147, 233
oil industry
  see also Deepwater Horizon oil spill
  alternative to natural gas 9, 56–57, 58,
    70–71
  consequences of exaggerated coverage
    155–158
  oceanic oil drilling 152, *152*
  oil spills 15, 20, 42–43, 150–153
  Santa Barbara project 153–154
Olivier, Darren 124

Pachauri, Rajenda 207
'peak oil' theory 56, 70–71
Pebble Bed Modular Reactor project 124
Pennsylvania Department of
  Environmental Protection 38, 61
Pennsylvania pollution 33, 37–38
Perciasepe, Robert 58
peregrine falcon 86
pesticides 83–89, 199
Pielke, Roger, Jr 205
Pilanesberg Game Reserve 253–254
Planet Green 141
poaching 246–249
polar bears 174–176, 255–259
polio 91
Politico Pro 60
pollution
  see also Deepwater Horizon oil spill
  coal 2, 37–38, 55–56, *56*, 58
  coal versus nuclear 128–129
  combustion emissions *56*
  misconceptions of pollution by natural
    gas 14, 23, 24–26, 27, 30–39, 40–41,
    55–62
  in Nigeria 42–43
  oil spills 15, 20, 42–43, 150–153
Population Bomb 229, 231
population growth 229–234

Power Balance bracelets 169
ProPublica 36
Pugh, Lewis Gordon 10–11, 12–21, 53

Rabi operation 155–156, 158
Radioactivity 48–52, *51*
  see also nuclear power
Raymond, Eric S. 203
Reddy, Christopher 148–149
Red List of Threatened Species (IUCN)
  246, 250, 255
Regis, Ed 231
Reiter, Paul 89
religion's role 223–227
renewable energy alternatives 56, 58, 71,
  76–77
rhinoceros 246–248, 249, 254–255
Richards, Zoe 127
Rose, Peter 65
rubella see MMR vaccine

Saier, James E. 205–206
Sands, Sarah 212
Santa Barbara project 153–154, 157
Sasol 12, 77
SASSI 218
Save Bantamsklip Association 120, 136
Schneider, Stephen 210–211
Schneiderman, Jill 143–144
Science 153, 204, 207
Scientific American 57, 128–129, 142, 145
Seale, Charly 251, 252
shale gas see fracking; natural gas
Shell
  fracking application in Karoo 11, 15,
    54, 55, 65, 222
  in Gabon 156
  in Nigeria 15, 20, 42–43
Silent Spring 82–83
Simon, Julian 219, 231–232, 241–242

Singh, Mayshree 74
Sirico, Robert 241
SKA project 63–67
Skone, Timothy 59
SkyTruth 62
Smallpox 90–91
Smithsonian Institution 156, 187
South African Sustainable Seafood
   Initiative (SASSI) 218
Späth, Andreas 39, 171
Square Kilometre Array *see* SKA project
Stockholm Convention (2001) 86
Stott, Phillip 237
Sturges, Tony 142
Suzuki Chemical Company 98

Taranto, James 176
Teaching Screens Production 123–124
Telegraph 147, 200, 237
TERI (The Energy and Resource
   Institute) 207
Thatcher, Margaret 236
The Age 190
The Gaia Hypothesis 211
The Great Global Warming Swindle 194, 236
The Hockey Stick Illusion 202
The Limits to Growth 231
The Nizkor Project 189–190
The Skeptical Environmentalist 229
The Times 175, 207
threatened species 128, 246, 255–256, 258
   see also endangered species
Three Mile Island nuclear power plant
   129, 130
Thyspunt 119
Tiger Bone and Rhino Horn 247
Tigers 248
TIME magazine 61, 149, 211
Tiplady, Adrian 64–65, 66, 75
TKAG

anti-fracking group 17, 21, 22
   exaggerated rhetoric 53, 78–80, 265
   propaganda 61–62, 70, 222
   sponsorship of journalist 39
Transocean 138–139
Treasure the Karoo Action Group
   see TKAG
Treehugger 141
Trouvelot, Étienne 87
Truth In Labeling Campaign 99, 100
't Sas-Rolfes, Michael 247
turtles 249–250
Twine, Tony 75–76, 264, 265

umami 97
UNEP 43, 45, 150, 151
United Nations Environment Programme
   see UNEP
United Nations Intergovernmental Panel
   on Climate Change *see* IPCC
United States
   see also Deepwater Horizon oil spill
   Barnett Shale 11, 63
   effects of DDT 82, 84–86, 87–88
   Environmental Protection Agency
      (EPA) 36, 58, 147, 154, 218
   fracking 9, 10, 11, 12, 32–34, 35, 38, 49,
      63, 69, 75
   Marcellus Shale 12, 38, 63, 69
   natural gas resources 63, 69
   Santa Barbara project 153–154, 157
University of East Anglia's CRU
   200–206
University of Maryland study 60
Uranium Road 123–125

vaccines 90–95
Vahrenholt, Fritz 208
Van Heerden, Ivor 147
Van Tonder, Gerrit 41

volcano (Iceland) 159
Von Liebig, Justus 97

Wakefield, Andrew 91–92, 93, 94, 95
Wall Street Journal 176
Walters, Stephen 239
Washington Post 217
Watermelons: The Green Movement's True Colors 237
Watts, Anthony 203
Weinberg, Robert 105
WHO 89, 99, 106, 130
Wigley, Tom 206
Wilson, Rob 203

Wired 231
Wood Mackenzie 60
Wood, Shauna 92
Woods Hole Oceanographic Institution 148
World Health Organization *see* WHO
World Wide Fund for Nature 11, 17, 120, 218, 257
Wyoming 35, 36, 38, 62–63

Yablokov, Alexey 131

Zeratsky, Katherine 99
Zorita, Eduardo 208–209

Do you have any comments, suggestions or
feedback about this book or any other Zebra Press titles?
Contact us at talkback@zebrapress.co.za

*

Visit www.randomstruik.co.za and subscribe
to our newsletter for monthly updates and news